T0258922

CRC SERIES IN ENZYME BIOLOGY

Editor-in-Chief
John R. Sabine, Ph.D.

3-HYDROXY-3-METHYLGLUTARYL COENZYME A REDUCTASE

Editor
John R. Sabine, Ph.D.
Department of Animal Sciences
Waite Agricultural Research Institute
University of Adelaide
Adelaide, Australia

PYRUVATE CARBOXYLASE

Editors
J. C. Wallace, Ph.D.
D. B. Keech, Ph.D.
Department of Biochemistry
University of Adelaide
Adelaide, Australia

CHOLESTEROL 7α-HYDROXYLASE (7α-MONOOXYGENASE)

Editors
Robin Fears, Ph.D.
Biosciences Research Center
Beecham Pharmaceuticals Research Division
Epsom, United Kingdom
John R. Sabine, Ph.D.
Department of Animal Sciences
Waite Agricultural Research Institute
University of Adelaide
Adelaide, Australia

PHOSPHATIDATE PHOSPHOHYDROLASE

Editor
David N. Brindley, Ph.D., D.Sc.
Lipid and Lipoprotein Research Group
Faculty of Medicine
University of Alberta
Edmonton, Canada

Phosphatidate Phosphohydrolase

Volume II

Editor

David N. Brindley, Ph.D., D.Sc.

Heritage Medical Scientist
Professor of Biochemistry
Lipid and Lipoprotein Group
Faculty of Medicine
University of Alberta
Edmonton, Canada

CRC Series in Enzyme Biology

Series Editor-in-Chief
John R. Sabine, Ph.D.

CRC Press
Taylor & Francis Group
Boca Raton London New York

CRC Press is an imprint of the
Taylor & Francis Group, an **informa** business

First published 1988 by CRC Press
Taylor & Francis Group
6000 Broken Sound Parkway NW, Suite 300
Boca Raton, FL 33487-2742

Reissued 2018 by CRC Press

© 1988 by Taylor & Francis
CRC Press is an imprint of Taylor & Francis Group, an Informa business

No claim to original U.S. Government works

This book contains information obtained from authentic and highly regarded sources. Reasonable efforts have been made to publish reliable data and information, but the author and publisher cannot assume responsibility for the validity of all materials or the consequences of their use. The authors and publishers have attempted to trace the copyright holders of all material reproduced in this publication and apologize to copyright holders if permission to publish in this form has not been obtained. If any copyright material has not been acknowledged please write and let us know so we may rectify in any future reprint.

Except as permitted under U.S. Copyright Law, no part of this book may be reprinted, reproduced, transmitted, or utilized in any form by any electronic, mechanical, or other means, now known or hereafter invented, including photocopying, microfilming, and recording, or in any information storage or retrieval system, without written permission from the publishers.

For permission to photocopy or use material electronically from this work, please access www.copyright.com (http://www.copyright.com/) or contact the Copyright Clearance Center, Inc. (CCC), 222 Rosewood Drive, Danvers, MA 01923, 978-750-8400. CCC is a not-for-profit organiza-tion that provides licenses and registration for a variety of users. For organizations that have been granted a photocopy license by the CCC, a separate system of payment has been arranged.

Trademark Notice: Product or corporate names may be trademarks or registered trademarks, and are used only for identification and explanation without intent to infringe.

A Library of Congress record exists under LC control number: 87020867

Publisher's Note
The publisher has gone to great lengths to ensure the quality of this reprint but points out that some imperfections in the original copies may be apparent.

Disclaimer
The publisher has made every effort to trace copyright holders and welcomes correspondence from those they have been unable to contact.

ISBN 13: 978-1-138-50578-0 (hbk)
ISBN 13: 978-1-138-56115-1 (pbk)
ISBN 13: 978-0-203-71111-8 (ebk)

Visit the Taylor & Francis Web site at http://www.taylorandfrancis.com and the CRC Press Web site at http://www.crcpress.com

SERIES PREFACE

The "CRC Series on Enzyme Biology" is a series of books, each one devoted to a single enzyme and each one endeavoring to draw together in a comprehensive and systematic manner all that is currently known about the chemistry, biochemistry, and physiology of that particular enzyme. Each volume, written or edited by one or more international enzyme specialists, draws together the latest information available on the occurrence, structure, function, role, and control of the biologically more important enzymes.

Each chapter or section of each volume is not written primarily for those specializing in that narrow area (innumerable specialist reviews do this), but rather to provide a coherent and integrated summary of that aspect for those working on other, quite different facets of the same enzyme. In this way several distinct classes of scientist, and student, will benefit from each volume, namely, those concerned with "type" situations of which that enzyme is an example, those working on the metabolic systems of which that enzyme is a key component, and those concerned with diseases in which the enzyme has an important role.

PREFACE

Phosphatidate phosphohydrolase is an enzyme that catalyzes a critical reaction in the synthesis of glycerolipids. The diacylglycerol that it produces is the precursor for the synthesis of triacylglycerols, phosphatidycholine, and phosphatidylethanolamine, and in plants for galactolipids. Consequently, its activity is essential in most tissues especially in order to provide the phospholipids that are needed for membrane formation. Because of this, no attempt has been made to review all papers on phosphatidate phosphohydrolase but rather we have selected tissues in which the enzyme has been characterized fairly extensively and for which there is a reasonable body of evidence concerning its role in metabolic regulation.

The mammalian tissues that have been chosen are liver, lung, and adipose tissue since their requirements for glycerolipid synthesis are specialized and fairly different. There is a further chapter that deals with plants and microorganisms. Each of these chapters has a general section that describes the special needs for glycerolipid synthesis and the physiological context in which the regulation of phosphatidate phosphohydrolase activity can be understood.

On a personal level, I should like to thank all of my colleagues who have worked with me on phosphatidate phosphohydrolase. I particularly wish to mention those who have offered criticism and advice in the preparation of these volumes and those whose work is illustrated in Chapters 1 and 2. These include: Mariana Bowley, Paul Bracken, Sue Burditt, Carmen Cascales, Maria Cascales, June Cooling, Robin Fears, Paul Hales, Roger Hopewell, Helen Glenny, Katherine Lloyd-Davies, Heather Mangiapane, Ashley Martin, Paloma Martin-Sanz, Janette Morgan, Sylva Pawson, Richard Pittner, Haydn Pritchard, Andy Salter, Janice Saxton, and Graham Sturton.

Finally, my deep gratitude to my Ph.D. supervisor and friend, the late Professor Georg Hubscher. He introduced me to phosphatidate phosphohydrolase and taught me so much about biochemistry.

SERIES EDITOR

John R. Sabine, M. Agr. Sc., Ph.D. is a Reader in Animal Physiology at the Waite Agricultural Research Institute of the University of Adelaide, Adelaide, Australia. Dr. Sabine obtained his Bachelor's and Master's degree in Agricultural Science from the University of Melbourne and then his Ph.D. in Animal Nutrition in the laboratory of Dr. B. Connor Johnson at the University of Illinois, Urbana. After several research appointments — Monash University, Australian National University, and the University of California, Berkeley, he was appointed to the faculty of University of Adelaide in 1967. At various times since then he has held Visiting Professorships at Brandeis University (Graduate Department of Biochemistry), the University of Stockholm (Wenner-Gren Institute), the University of Oklahoma Health Sciences Center (Department of Biochemistry and Molecular Biology), and the University of Kuwait (Department of Biochemistry), as well as Visiting Scholar appointments at Oxford University (Department of Clinical Medicine) and Harvard University (Chemistry Department).

Dr. Sabine is a member of several scientific societies, including particularly the Australian Biochemical Society, the Nutrition Society of Australia, and the Australian Physiological and Pharmacological Society. He has presented his research findings at various national and international meetings, and has been chairman of a number of sessions at these meetings. He was co-convenor of, and leader of the Australian delegation to, the U.S./Japan/Australia Cancer Conference (Hawaii, 1975) and the convenor and chairman of the unique international symposium "Lipids in Cancer", which was held on board the Indian-Pacific Express as a satellite meeting to the 12th International Congress of Biochemistry, 1982. In 1979 he delivered the 12th Patricia Chomley Oration to the Australian College of Nursing.

Dr. Sabine has published some 50 research articles and 14 invited reviews. He is on the Editorial Board for *Nutrition and Cancer*, and his book *Cholesterol* was the first comprehensive coverage of this important field to appear in 20 years. His major research interests revolve around cholesterol physiology, with particular reference to the control of its synthesis and to its role in membrane structure and function, and with emphasis upon the role of cholesterol in the etiology of cancer. He has further research interests in such diverse fields as the role of earthworms in biological resource recovery, the physiology of goats for meat and fiber production and the interaction between scientists and society.

THE EDITOR

David N. Brindley, Ph.D., D.Sc. was Professor of Metabolic Control in the Department of Biochemistry at the University of Nottingham, Nottingham, England.

Professor Brindley received his undergraduate training in the Department of Medical Biochemistry, University of Birmingham and received a B.Sc. (1st Class Honours) in 1963. He then studied for a Ph.D. in the same Department under the supervision of the late Professor G. Hübscher and gained his Ph.D. in 1966. A further year was then spent as a Postdoctoral Fellow with Professor Hübscher. The work was concerned with the control of glycerolipid synthesis in the small intestine in relation to fat absorption.

In 1967 Professor Brindley moved to Harvard University where he worked for two years as a Postdoctoral Fellow for Professor K. Bloch. The work investigated the synthesis of fatty acids in *Mycobacterium phlei*. A multienzyme complex of fatty acid synthetase was shown to occur in a bacterium for the first time and novel stimulating factors for fatty acid synthesis were demonstrated.

Professor Brindley retruned to England in 1969 to join the newly established Medical School in Nottingham. He was subsequently promoted from Lecturer to Senior Lecturer, Reader, and then Professor and he gained his D.Sc. from the University of Birmingham in 1977. From January 1, 1988, he has accepted the position of Heritage Medical Scientist and Professor of Biochemistry in the newly formed Lipid and Lipoprotein Research Group that is sponsored by the Alberta Heritage Foundation for Medical Research.

His main interests are the effects of hormones, metabolites, drugs, and diet in regulating: (a) glycerolipid synthesis particularly at the level of phosphatidate phosphohydrolase; (b) the secretion of very low density lipoprotein and lysophosphatidycholine from the liver; (c) the binding and uptake of low and high density lipoproteins by the liver; (d) insulin responsiveness and metabolism in adipose tissue; and (e) food intake and the level of circulating glucose and triacylglycerol.

CONTRIBUTORS

David N. Brindley, Ph.D., D.Sc.*
Heritage Medical Scientist
Professor of Biochemistry
Lipid and Lipoprotein Group
Faculty of Medicine
University of Alberta
Edmonton, Canada

John L. Harwood, Ph.D., D.Sc.
Professor
Department of Biochemistry
University College
Cardiff, Wales

Fred Possmayer, Ph.D.
Professor
Department of Obstetrics and Gynecology,
  and Biochemistry
University of Western Ontario
London, Ontario, Canada

Molly J. Price-Jones, Ph.D.
Research
Department of Biochemistry
University College
Cardiff, Wales

E. D. Saggerson, Ph.D., D.Sc.
Reader in Biochemistry
Biochemistry Department
University College London
London, England

* At the time these volumes were written, Dr. Brindley was Professor of Metabolic Control, Department of Biochemistry, University of Nottingham, Nottingham, England.

TABLE OF CONTENTS

Volume I

Volume II

Chapter 4

PHOSPHATIDATE PHOSPHOHYDROLASE IN PLANTS AND MICROORGANISMS

J. L. Harwood and M. J. Price-Jones

TABLE OF CONTENTS

I. PLANT LIPIDS AND THE ROLE OF PHOSPHATIDATE PHOSPHOHYDROLASE IN THEIR SYNTHESIS

Phosphatidate phosphohydrolase plays a central role in lipid metabolism in plants. Its importance in individual pathways in different plants varies (as will be seen later) but, in general, the diacylglycerol it produces is utilized both for membrane lipid formation and for storage triacylglycerol production. The major nonchloroplastic lipids, phosphatidylcholine and phosphatidylethanolamine, are made primarily by the CDP-base pathway which uses diacylglycerol in its final step.[1,2] Furthermore, diacylglycerol is also the acceptor for galactose in the synthesis of the major chloroplast lipid, monogalactosyldiacylglycerol.[3] Even when one considers other acyl lipids, such as phosphatidylglycerol, whose formation does not depend on diacylglycerol, the activity of phosphatidate phosphohydrolase is important because of the competition for phosphatidate substrate.

Pathways of glycolipid formation in plants are shown in Figure 1 and the central roles of phosphatidic acid and phosphatidate phosphohydrolase are emphasized in Figure 2. One complication which is not brought out adequately in these diagrams is that the various substrates for specific reactions are not always generated in the same part of the cell. This fact is coupled with particular and often different subcellular distributions for the individual enzymes. In short, for the efficient production of most glycerolipids in plants there must be a cooperation between organelles. We shall, therefore, discuss first present ideas for fatty acid, phosphoglyceride, and glycosylglyceride synthesis in order to place phosphatidate phosphohydrolase in its proper subcellular context. First of all though, a few brief remarks about the nature of plant lipids will be appropriate since these differ in several respects from mammalian lipids.[4,5]

Plant membrane lipids are generally much more unsaturated than mammalian lipids (Table 1). This is especially true of photosynthetic membranes. A typical analysis of leaf tissue would show 10% linoleate and 60% α-linolenate. In most storage triacylglycerols, linoleate is a major component (e.g., 60% soybean oil, 52% maize oil) although seed oils show much more diversity in composition than membrane fractions. The fatty acid composition of individual lipid classes is usually quite characteristic.[5,6] In particular, the occurrence of the unusual *trans*-Δ3-hexadecenoate is confined to phosphatidylglycerol and all *cis*-Δ7,10,13-hexadecatrienoate ("16:3") is found in the monogalactosyldiacylglycerol of some plants. The special role of phosphatidate phosphohydrolase in 16:3 vs. 18:3 plants is discussed in Section IV. The highly unsaturated nature of plant lipids is also maintained in the marine environment, although marine algae accumulate generally 20C polyenoic acids such as arachidonic (all *cis*-Δ5,8,11,14-eicosatetraenoic acid) or all *cis*-Δ5,8,11,14,17-eicosapentaenoic acid in excess of 18C unsaturated acids.

Photosynthetic membranes from all O_2-evolving organisms (higher plants, mosses, algae, cyanobacteria) possess a unique and ubiquitous lipid pattern[5] (Table 1). The major constituent, accounting for 45 to 50% of the total acyl lipids, is monogalactosyldiacylglycerol. The next most prevalent lipid is digalactosyldiacylglycerol (25 to 30%). A particular sulfur-containing lipid, sulfoquinovosyldiacylglycerol, accounts for about 10% of the total and the only significant phospholipid is phosphatidylglycerol (about 10%). Recent work has shown that the two envelope membranes of higher plant chloroplasts have quite different compositions. While the inner envelope membrane resembles the photosynthetic thylakoids, the outer envelope membrane has a composition similar to the endoplasmic reticulum.[7,8]

Extra-chloroplastic membranes in algae and higher plants seem to resemble those of animals in that phosphatidylcholine and phosphatidylethanolamine are major lipids (Table 1), phosphatidylinositol is a significant component,[5] and diphosphatidylglycerol is confined to the inner mitochondrial membrane.[9] Marine algae seem to have a more complex distribution of membrane lipids, but because there has been little analysis to date, it is difficult to make generalizations except that various glycolipids seem to be significant constituents.

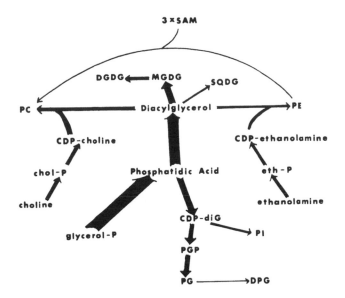

FIGURE 1. Pathways of polar glycerolipid synthesis in higher plant leaves. The thickness of the arrows represents the relative importance of individual steps in leaf tissues.

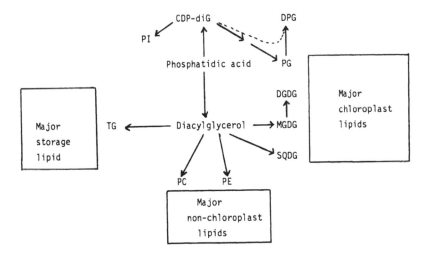

FIGURE 2. The central role of phosphatidate phosphohydrolase in glycerolipid metabolism in plant tissues. The exact pathway for diphosphatidylglycerol formation has not been demonstrated and is indicated by a dotted line. Abbreviations: PI, phosphatidylinositol; CDP-diG, CDP-diacylglycerol, DPG, diphosphatidylglycerol; PG, phosphatidylglycerol; TG, triacylglycerol; PC, phosphatidylcholine; PE, phosphatidylethanolamine; MGDG, monogalactosyldiacylglycerol; DGDG, digalactosyldiacylglycerol; SQDG, sulfoquinovosyldiacylglycerol.

The high levels of polyunsaturated fatty acids in algal and plant membranes have already been mentioned. This fact makes the activity of the various aerobic desaturases particularly relevant although we know much less about these enzymes than about, for example, fatty acid synthase. The general process of fatty acid synthesis in plants (Figure 3) involves *de novo* synthesis of palmitate by fatty acid synthase, elongation to stearate, and Δ9-desaturation to oleate. These processes involve generally water-soluble acyl-ACPs (acyl-acyl carrier proteins). Further desaturation at the 18C level utilizes complex lipids as substrates.[10] The

Table 1

GENERAL FEATURES OF PLANT AND ALGAL ACYL LIPIDS

Fatty Acids

16:0, 18:1, 18:2, α-18:3	Major components of membrane lipids of terrestrial plants/freshwater algae
16:0, 18:1, 18:2	Major acids of storage fats (sometimes unusual or minor acids are major components)
16:0, 18:1, 20:4, 20:5	Major acids in membrane lipids of marine algae

Photosynthetic Membranes

Monogalactosyldiacylglycerol (50%), digalactosyldiacylglycerol (25%), and sulfoquinovosyldiacylglycerol (10%) are major lipids. Phosphatidylglycerol (10%) is only important phosphoglyceride.
  All the above lipids have very specific fatty acid distributions.

Nonphotosynthetic Membranes

Phosphatidylcholine and phosphatidylethanolamine are major constituents in terrestrial plants and all algae except blue-greens (cyanobacteria). Glycolipids and various uncharacterized lipids occur as well as phosphoglycerides in marine algae.

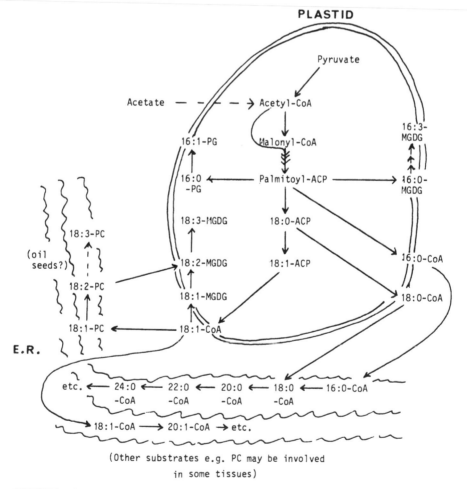

FIGURE 3.   Fatty acid synthesis in plants. The possible contribution of acetate to acetyl-CoA availability and of phosphatidylcholine for linoleate desaturation in oil seeds remains to be assessed quantitatively. Abbreviations as for Figure 2 and ACP, acyl carrier protein.

generation of palmitate and oleate is a general property of a chloroplastic type II synthetase, a specific β-ketoacyl ACP synthetase,[11] and a Δ9-desaturase.[12] Further modifications of the acyl chains can take place within the plastid (desaturation) or in the extra-chloroplastic compartment. Indirect evidence implies that formation of *trans*-Δ3-hexadecenoic acid takes place while palmitate is esterified to the *sn*-2 position of phosphatidylglycerol while hexadecatrienoate synthesis is likely to occur while in the intact monogalactosyldiacylglycerol molecule.[13]

In contrast, desaturation of oleate to linoleate involves a circuitous route. Oleoyl-ACP in the stroma is hydrolyzed and oleate is converted to oleoyl-CoA in the envelope.[14,15] This oleoyl-CoA can then be incorporated into phosphatidylcholine via monoacylphosphatidylcholine (see Sections IV and V). Phosphatidylcholine appears to be the substrate for oleate desaturation[13] though whether this takes place in endoplasmic reticulum or whether the Δ12-desaturase is present (also) in the outer envelope of plastids is not known. Further desaturation to linolenate may take place on phosphatidylcholine during seed oil formation but there is good evidence that monogalactosyldiacylglycerol is the substrate in leaf tissues.[13] It was shown in labeling experiments with leaf tissues that newly formed linolenate was associated with this lipid[16,17] and a sequential disappearance of linoleate and appearance of linolenate was found in the monogalactosyldiacylglycerol of isolated chloroplasts.[18] A direct demonstration of desaturation of monogalactosyldiacylglycerol by chloroplasts in vitro confirmed these conclusions.[19] Furthermore, the problem of how linoleate in extra-chloroplastic phosphatidylcholine could be moved to chloroplastic monogalactosyldiacylglycerol was solved by the demonstration that phospholipid exchange proteins would allow linoleoyl-phosphatidylcholine to be desaturated by isolated chloroplasts[19,20] but only after incorporation of the acyl chain into monogalactosyldiacylglycerol.[20] Monogalactosyldiacylglycerol is also important for fatty acid desaturation in cyanobacteria.[21]

Elongation of long chain fatty acids and further modifications of acyl chains probably take place outside the plastid. Indeed, Stumpf et al. have summarized available evidence into a two-compartment model.[22] Thus, for most major fatty acids, more than one compartment is involved in their synthesis.

Synthesis of phosphoglycerides can be divided into the CDP-diacylglycerol (phosphatidylinositol, phosphatidylglycerol, diphosphatidylglycerol) and the diacylglycerol (phosphatidylethanolamine, phosphatidylcholine) pathways (Figure 1). These two key intermediates are both generated from phosphatidic acid which lies thus at a branch point in metabolism. When they have been tested, the pathways for phosphoglyceride synthesis appear to be the same in all plants with the possible exception of phosphatidylserine which is formed via CDP-diacylglycerol in spinach[23] but by an exchange reaction in castor bean.[2] Although the endoplasmic reticulum is the primary site of phospholipid synthesis in higher plants, significant amounts of many reactions have been measured in other fractions[2] (Figure 4). For example, mitochondria are capable of synthesizing phosphatidic acid, CDP-diacylglycerol, phosphatidylcholine (by both the methylation and CDP-base pathways), phosphatidylethanolamine, phosphatidylglycerol, and diphosphatidylglycerol.[9] Some of these reactions (e.g., phosphatidic acid and phosphatidylglycerol synthesis) are at comparable levels in mitochondrial and endoplasmic reticulum fractions while others are mainly confined to endoplasmic reticulum.[2] The major (often only) phospholipid in plastids is phosphatidylglycerol; chloroplasts have been shown to be capable of its synthesis,[24] the activity being present in envelope membranes.[25] Movement of phosphoglycerides from their site of synthesis to a final location presumably takes place by intramembrane flow or by the use of one of the phospholipid exchange proteins isolated from plant tissues.[26]

Monogalactosyldiacylglycerol and digalactosyldiacylglycerol are formed by the sequential galactosylation of diacylglycerol.[3] The source of UDP-galactose is the cytosol[27] and the diacylglycerol can either be generated by plastid phosphatidate phosphohydrolase or mainly

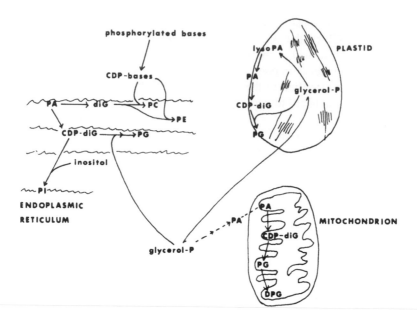

FIGURE 4.   Contribution of different subcellular sites to phosphoglyceride synthesis in plants. Abbreviations as for Figure 2 and PA, phosphatidic acid.

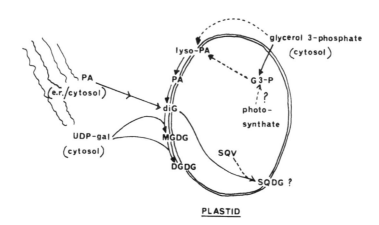

FIGURE 5.   Contribution of different subcellular sites for glycosylglyceride synthesis in plants. Abbreviations as in Figure 2 and G 3-P, glycerol 3-phosphate; PA, phosphatidic acid; UDP-gal, UDP-galactose; SQV, sulfoquinovose.

from outside the plastid (see discussion on 16:3 and 18:3 plants in Section IV). Higher galactolipid homologs can also be generated by galactolipid:galactolipid galactosyltransferase.[28] This enzyme would also provide a source of diacylglycerol in chloroplasts.

The synthesis of the third major glycolipid, sulfoquinovosyldiacylglycerol, remains an enigma.[29] However, recent studies have managed to demonstrate the expected formation of this sulfolipid in chloroplasts.[30] Again the diacylglycerol which is, presumably, necessary for its synthesis[29] can come from both internal and external plastid sources depending on the plant type.[30a] These interactions for glycolipid formation are summarized in Figure 5.

It will be clear from the foregoing that, not only do plants contain large amounts of lipid which are often absent or only minor in animals but, in addition, the synthesis of these lipids involves a considerable amount of compartmentation and interaction of organelles.

Table 2
**TIME COURSE OF THE INCORPORATION OF
RADIOACTIVITY FROM ³²P-ORTHOPHOSPHATE INTO
THE PHOSPHOLIPIDS OF DEVELOPING SOYBEAN
SEEDS OR FRONDS FROM THE MARINE BROWN
ALGA *Fucus Serratus***

Lipid	Percent total lipid radio-activity (hr)				
	1/4	1	2	4	24
Soybean[32]					
Phosphatidic acid	58	33	11	8	5
Phospholipids formed mainly via CDP-bases[a]	16	37	53	60	67
Phospholipids formed mainly via CDP-diG[b]	17	28	34	31	27
F. serratus[35]					
Phosphatidic acid	—	14	10	7	2
Phospholipids formed via CDP-bases[a]	—	51	50	55	60
Phospholipids formed via CDP-diG[b]	—	35	40	38	38

[a] Includes PC, PE, and *N*-acyl PE (soybean).
[b] Includes PG, DPG, and PI.

II. LABELING STUDIES IN DIFFERENT PLANT OR ALGAL TISSUES

Evidence for the activity of phosphatidate phosphohydrolase has come indirectly from two sorts of experiments. First, the dynamic state of the phosphatidic acid molecule has been shown during ³²P-orthophosphate incorporations. Second, and more importantly, the metabolism of diacylglycerol (and phosphatidic acid) has been followed in double labeling experiments employing, for example, [2-³H]glycerol and [1-¹⁴C]acetate. These will be dealt with in turn.

Incorporation of radioactivity from ³²P-orthophosphate into phospholipids during time-course experiments usually confirms that phosphatidic acid is labeled at high rates initially. This high specific activity then rapidly declines as phosphatidate is dephosphorylated or converted to CDP-diacylglycerol. Two examples of such a set of results are shown in Table 2. Soybean tissue has been quite well studied with regard to its metabolism and it is known that phosphatidylcholine is mainly made by the CDP-base pathway.[31] Thus, it can be assumed that a significant proportion of the decline in phosphatidate labeling is due to phosphatidate phosphohydrolase activity which produces the necessary diacylglycerol. (It should be borne in mind that the ³²P-labeling of phosphatidylcholine and phosphatidylethanolamine will be via ATP whereas phosphatidylinositol or phosphatidylglycerol will be labeled from phosphatidate which would itself be labeled either from ATP or from other sources such as glycerophosphate). The rapid and transient labeling of phosphatidic acid in maturing soybean[32] was also seen in studies with other plants or algae. Data for the marine brown alga, *Fucus serratus* are included in Table 2 and Bieleski[33] showed a similar result with *Spirodela oligorhiza* L.

The labeling experiments with soybean were continued by Wilson and Rinne[34] who used pulse labels from [¹⁴C]acetate and [2-¹⁴C]glycerol. Although their studies are not simple to interpret, the data again indicated a very rapid turnover and degradation of phosphatidate.

The high rates of labeling of diacylglycerol and, after prolonged incubations, of triacylglycerol, suggested that a large proportion of the degradation of phosphatidate resulted from phosphohydrolase action to yield the necessary diacylglycerol intermediate. The use of appropriate radiolabeled precursors is very important and it has been shown for a number of tissues that phosphoglycerides are labeled in their acyl moieties from [1-^{14}C]acetate[16] whereas [2-^{3}H]glycerol only labels the glycerol moiety of the phospholipid. This is because the radiolabel is lost from glycerol 1-phosphate when it is converted to dihydroxyacetone phosphate which eventually gives rise to acetyl-CoA the precursor of fatty acids.[36]

Labeling studies with various precursors have given information on the generation of diacylglycerol for galactolipid and/or phosphoglyceride synthesis. Much of this work has been thoroughly reviewed and discussed by Roughan and Slack[13] and only a summary will be made here. The cooperation of organelles in the formation of complex lipids was discussed in Section I and the labeling studies should be seen in that context. Almost invariably in overall labeling studies with leaves or leaf disks, radioactivity accumulates to the greatest extent in phosphatidylcholine. Depending on the tissue, and also what developmental stage it was in,[17,37] then monogalactosyldiacylglycerol could also be highly labeled. These two lipids both utilize diacylglycerol generated by phosphatidate phosphohydrolase (Figures 1, 4, and 5). However, although galactolipid synthesis is confined to the plastid[3] and UDP-galactose:diacylglycerol galactosyltransferase is located on the inner chloroplast envelope,[8] labeling of phosphatidylcholine indicated an extra-chloroplastic site for its synthesis. Thus, highly labeled phosphatidylcholine synthesized from [^{14}C]acetate by expanding pea (*Pisum sativum* L.) leaves was localized by sucrose-density gradient centrifugation of leaf homogenates predominantly within a band corresponding to microsomes.[38] Relatively little radioactivity was associated with the phosphatidylcholine of intact chloroplasts — this phosphoglyceride being a significant component of the outer envelope membrane where it is concentrated in the cytoplasmic leaflet.[39,40] Similar studies with other tissues and/or different precursors have confirmed that the major site of incorporation of newly synthesized fatty acids (particularly oleate) into phosphatidylcholine lies outside the chloroplast.[13]

When these types of experiments were repeated with [2-^{3}H]glycerol the results were consistent with the proposal that the whole diacylglycerol moiety and not just the fatty acids of phosphatidylcholine could be utilized for chloroplastic galactolipid synthesis. Examination of the ^{14}C/^{3}H ratios in double labeling ([1-^{14}C]acetate + [2-^{3}H]glycerol) experiments suggested that labeled fatty acids were incorporated into phosphatidylcholine as a result of net synthesis via phosphatidic acid and that diacylglycerol derived from phosphatidylcholine was then used for monogalactosyldiacylglycerol formation. Furthermore, the ratios originally present within the oleate- and linoleate-containing molecular species of phosphatidylcholine were retained within the newly synthesized monogalactosyldiacylglycerol.[13] The turnover of phosphatidylcholine seems to be particularly rapid when cells are actively forming galactolipids (such as in chloroplast development) or when they are accumulating triacylglycerol.

Because plastids seem to be the major site for both *de novo* fatty acid synthesis and chloroplastic lipid (monogalactosyldiacylglycerol, digalactosyldiacylglycerol, sulfoquinovosyldiacylglycerol, phosphatidylglycerol) formation, several laboratories studied the incorporation of radioactivity from suitable precursors into different lipid classes. These experiments have demonstrated firmly the presence of an active chloroplast phosphatidate phosphohydrolase (see Section III) which can generate a pool of freshly synthesized diacylglycerol for further metabolism. In a typical incubation of spinach chloroplasts with [1-^{14}C]acetate the products were mainly oleate and palmitate with only small amounts of polyunsaturated fatty acids. The [^{14}C]fatty acids were distributed within four major classes: nonesterified fatty acids (70 to 80%), acyl thioesters (<5%), glycerolipids (10 to 12%), and diacylglycerol (10 to 15%).[13,41] Addition of Triton X-100 increased the accumulation of label in diacylglycerol. However, this compound did not alter the relative amounts of

[^{14}C]oleate and [^{14}C]palmitate formed. In contrast, glycerol 3-phosphate not only stimulated the amounts of diacylglycerol accumulated but also increased the relative amount of palmitate at the expense of oleate.[41] At low concentrations of glycerol 3-phosphate (50 to 100 μM) the main product of phosphatidate phosphohydrolase action was 1-oleoyl, 2-palmitoyl diacylglycerol whereas at high concentrations (> 1 mM) significant quantities of dipalmitoyl-diacylglycerol were found. The acylation of glycerol 3-phosphate is a stepwise process. A soluble glycerol 3-phosphate acyl transferase[42] and an envelope-bound monoacylglycerol-3-phosphate acyltransferase[43] are responsible for the two acylation reactions to produce phosphatidate. The former enzyme can use both acyl-CoA and acyl-ACP substrates for the acylation of the C-1 position of glycerol 3-phosphate and prefers oleate.[44] The monoacylglycerol 3-phosphate acyltransferase preferentially utilizes ACP-esters and palmitate rather than oleate.[45] By the combined action of these two acyltransferases, phosphatidic acid is produced which has the 1-oleoyl, 2-palmitoyl structure. The speed with which this phosphatidate is then dephosphorylated to yield chloroplastic diacylglycerol for incorporation into galactolipids depends on the plant type (see Section IV).

The molecular species of the chloroplastic lipids, phosphatidylglycerol and sulfoquinovosyldiacylglycerol, which are formed in labeling studies also resemble those of the accumulated diacylglycerol in isolated chloroplasts.[46] These results again emphasize the pivotal role of phosphatidic acid metabolism not only in determining the overall proportions of different lipid classes but also in influencing exact molecular species distributions. Although earlier interpretation of data proposed that cytoplasmic phosphatidic acid was used for CDP-diacylglycerol formation and phosphatidylglycerol synthesis[13] recent studies have demonstrated that plastids can make the latter phospholipid. Indeed, both CDP-diacylglycerol and phosphatidylglycerol synthesis are found on the envelope membranes where phosphatidic acid formation takes place.[25] Thus, it is clear that the phosphatidate phosphohydrolase of chloroplasts has to compete with the CTP:phosphatidic acid cytidylyltransferase for available phosphatidate substrate. In that connection it is noteworthy that the cytidylyltransferase is severely inhibited by Triton X-100[25] and, therefore, the increase in diacylglycerol labeling in intact chloroplasts with this detergent[13] could have been due to the inhibition of competitive side reactions for phosphatidate phosphohydrolase.

Labeling studies with leaf systems have also given information about the presumed extra-chloroplastic metabolism of phosphatidic acid. The final steps in the synthesis of the major extra-chloroplastic lipids, phosphatidylcholine and phosphatidylethanolamine and also phosphatidylinositol, take place primarily on the endoplasmic reticulum.[1,2] Thus, once more, the phosphatidate is utilized by both a phosphatidate phosphohydrolase and a cytidylyltransferase. Acylation of cytoplasmic glycerol 3-phosphate uses acyl-CoAs. There is evidence from labeling studies in a variety of tissues that, in contrast to the chloroplast, palmitate is transferred specifically to position 1 while oleate can be transferred to both positions 1 and 2 of glycerol 3-phosphate. This results in the formation of both dioleoyl- and 1-palmitoyl, 2-oleoyl-phosphatidic acid. While phosphatidate phosphohydrolase appears to utilize both species the cytidylyltransferase appears to prefer the 1-palmitoyl, 2-oleoyl-species[13] thus accounting for the enrichment of palmitate at the 1-position of phosphatidylinositol.[47]

The marked effect that exogenous glycerol 3-phosphate had on the relative production of diacylglycerol by phosphatidate phosphohydrolase in isolated chloroplasts (above) can provide an explanation for earlier labeling studies. It is known that glycerol 3-phosphate synthesis is localized in the cytosol[27] and, therefore, its cytoplasmic concentration should produce a similar effect in vivo. The observation that the physiological status of pumpkin leaves influenced the relative labeling of different lipids can be explained by increased concentrations of glycerol 3-phosphate in those leaves where relative labeling of phosphatidylcholine was decreased. Thus, newly synthesized fatty acids were chaneled towards chloroplastic phosphatidylglycerol.[13] However, little is known about other possible control factors in vivo.

III. PROPERTIES OF DIFFERENT PLANT PHOSPHATIDATE PHOSPHOHYDROLASES

The first demonstration of phosphatidate phosphohydrolase activity in subcellular fractions from plants was by Kates.[48] He examined the hydrolysis of exogenous phosphatidylcholine by crude plastid fractions from cabbage, spinach, and sugar beet leaves and carrot root. Choline was liberated initially from the substrate in a first order reaction due to phospholipase D action and, subsequently, the phosphatidate produced was shown to be hydrolyzed with the liberation of inorganic phosphate. As might be expected the phosphatidate phosphohydrolase reaction showed a short lag phase. It was also relatively slow compared to the phospholipase D reaction. Quantification of the phosphatidate phosphohydrolase activity was complicated because it was not measured directly and because phospholipase A activity in the same extracts gave rise to monoacylphosphatidate and glycerophosphate from phosphatidate and also because the glycerophosphate itself was hydrolyzed in the system to yield inorganic phosphate.

Nevertheless, some characteristics of phosphatidate phosphohydrolase in these fractions could be deduced. Enzyme activity was measured in buffers in the pH range 4.5 to 7.5 without any detergent and using egg yolk phosphatidylcholine as substrate. Greatest liberation of inorganic phosphate was found for the spinach chloroplast fraction (70% of available phosphate). Sugar beet and cabbage chloroplast fractions contained low activity (12 and 5% of available phosphate) but carrot root chromoplasts, in spite of a very active phospholipase D, did not liberate inorganic phosphate. Since phosphatidate phosphohydrolase was not measured directly and because glycerophosphate was hydrolyzed at twice the rate of phosphatidate in spinach,[48] it is difficult to be certain how active the spinach phosphatidate phosphohydrolase was in these experiments. However, the pH optimum of inorganic phosphate release (pH 5.0) was somewhat lower than that for glycerophosphate hydrolysis (pH 5.2) and this, together with the relative amounts of intermediates found, suggested that phosphatidate phosphatase made a major contribution to the inorganic phosphate released. Monoacylphosphatidate and cholinephosphate were not hydrolyzed. Perhaps the most significant observations by Kates were the very similar (acid) pH optima for the liberation of choline and inorganic phosphate by the spinach chloroplast fraction. In view of the later demonstration of an alkaline phosphatidate phosphohydrolase in spinach chloroplasts (see below), this activity may reflect the availability of substrate phosphatidate generated by phospholipase D.[48]

An acid phosphatase from wheat germ which was normally assayed using O-carboxyphenyl phosphate as substrate was found to have activity towards phosphatidate.[49] A beef liver phosphatidylcholine was used to prepare the phosphatidic acid substrate which was incubated in the absence of detergent at pH 5.9 (no pH curve was made). The acid phosphatase showed comparable activity with phosphatidic acid, 1-monoacylphosphatidate, and 1-alkylglycerol-3-phosphate substrates. Less activity was found with 1-alkyl-2-acylglycerol-3-phosphate and 1-alkyldihydroxyacetone phosphate substrates and no activity with phosphatidylcholine.[49] These results raise the possibility that some phosphatidic acid in plants can be degraded by a nonspecific acid phosphatase rather than by a specific phosphatidate phosphohydrolase. Konigs and Heinz[50] also found phosphatidate phosphohydrolase activity with an acidic pH optimum. They studied the hydrolysis of phosphatidic acid prepared from a plant source (with a much higher double bond index than egg lipid) in the presence of the nonionic detergent Triton X-100. The substrate was also dispersed by sonication. Perhaps, as a result of these methods, activity in their broad bean (*Vicia faba* L.) extracts was relatively high and they were able to use short (1 min) incubations. Phosphatidate phosphohydrolase activity was found in two subcellular fractions. One was sedimentable at 15,000 g and this membrane-bound activity could not be made soluble by sonication but could be released by detergent

Table 3
SUBCELLULAR LOCATION AND pH OPTIMA OF VARIOUS
PLANT PHOSPHATIDATE PHOSPHOHYDROLASES

Tissue	Subcellular localization	pH Optimum	Notes
Spinach[48]	Crude chloroplast fraction	5.0	Assayed indirectly
Vicia faba[50]	15,000-g particulate	5.6	Enzyme very unstable
	40,000-g supernatant	5.1	
Castor beans (germinating)[51]	Microsomal		Assayed at pH 7.4
Mung bean cotyledons[53]	Endoplasmic reticulum	7.5	Some acidic pH activity
	Soluble	5.0	associated with protein bodies
Wheat germ[49]	Soluble preparation		Assayed at pH 5.9 Nonspecific
Spinach[43,57]	Chloroplast envelope	9.0	Measured indirectly via G3P incorporation

treatment. This particulate fraction was further purified by sucrose density-gradient centrifugation but the procedure caused a loss of phosphatidate phosphohydrolase activity. Differential centrifugation showed that the particulate enzyme was present in a fraction containing both mitochondria and microbodies but was not found in significant amounts in either chloroplastic or microsomal fractions (Table 3). The particulate activity had a pH optimum of 5.6, an apparent K_m for phosphatidic acid of 3.5 μm, and a specific activity of 15 to 20 nmol/min/mg protein. The activity was sensitive to Ca^{2+}, 10 μm Ca^{2+} causing 50% inhibition, and 100 μm Ca^{2+} producing complete inhibition. The enzyme was also very unstable.

A soluble phosphatidate phosphohydrolase was also found in *Vicia faba* leaves.[50] This enzyme had comparable activity (8 to 15 nmol/min/mg protein) to the particulate enzyme, but a lower pH optimum (5.1) and was less sensitive to Ca^{2+} inhibition (K_i = 100 μm).

The activities of these two phosphatidate phosphohydrolases were studied in greening etiolated seedlings of *V. faba*. The results are compared with the appearance of total chlorophyll in Figure 6. Both activities were present in etiolated seedlings and caused a rapid increase in both so that maximal activities were found after 1 day. The soluble activity then declined somewhat while the particulate phosphatidate phosphohydrolase remained constant. In contrast, chlorophyll accumulation (and chloroplast development) occurred at significantly later times.[50] However, these workers were careful to point out that the lack of sensitivity of their assay may have prevented their detecting chloroplastic phosphatidate phosphohydrolase particularly in the subcellular fractionations.

In contrast to the failure of Konigs and Heinz[50] to detect any phosphatidate phosphohydrolase in microsomal fractions from *V. faba* L., Moore et al.[51] using germinating castor beans concluded that this fraction was the major site of particulate activity. However, in their assay they used 4 mM phosphatidic acid (three orders of magnitude greater than the K_m for the *V. faba* enzyme[50]) which was obtained from an animal source, no detergent, and an assay pH of 7.4. Neither glucose 6-phosphate nor *O*-carboxyphenyl phosphate were hydrolyzed in their assays and the activity was sensitive to boiling. No activity was found associated with either the mitochondrial or the glyoxysomal bands on sucrose density gradient centrifugation. Also in contrast to the *Vicia faba* particulate activity which was only linear for 1 min, the castor bean enzyme liberated phosphate in a linear fashion for at least 60 min. Activities of the order of 1.3 nmol/min/mg protein were found for the microsomal fraction. In addition to the microsomal activity, Moore et al.[51] reported that about one third of the total recovered phosphatidate phosphohydrolase activity in castor bean was soluble. No further characterization of these two fractions was carried out but Moore and Sexton[52]

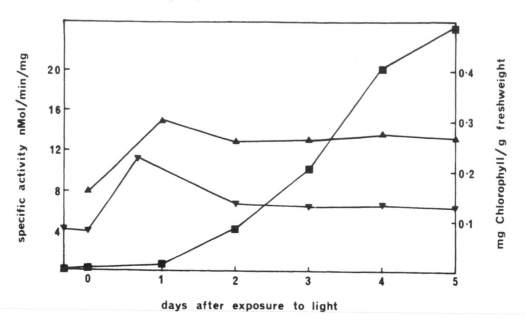

FIGURE 6.   Development of phosphatidate phosphohydrolase activity in greening etiolated seedlings of *Vicia faba* L. (■—■) Chlorophyll, (▼—▼) membrane-bound phosphatidate phosphohydrolase, (▲—▲) soluble phosphatidate phosphohydrolase.[50]

also examined the soluble enzyme from castor bean. They found that the soluble fraction had peaks of activity at pH 4.25 and 6.0 with apparent K_m's of 4.0 and 1.75 mM respectively. They recalculated the contribution of soluble and microsomal activities and concluded that, when these fractions were assayed under optimal conditions, there was 12-fold more soluble phosphatidate phosphohydrolase than total endoplasmic reticulum activity.

The phosphatidate phosphohydrolase from mung bean cotyledons was localized in subcellular fractions prepared by isopycnic sucrose density separation of homogenates.[53] The fractions were assayed at both pH 7.5 and 5.0. Bands corresponding to plasma membrane, Golgi apparatus, mitochondria, endoplasmic reticulum, and protein bodies were formed and identified with marker enzymes. Activity of phosphatidate phosphohydrolase assayed at pH 7.5 was localized in the same fractions as NADH-cytochrome c reductase. Thus, in agreement with the castor bean work, the pH 7.5 phosphatase was probably present in the endoplasmic reticulum. In contrast, when assayed at pH 5.0 phosphatidate phosphohydrolase activity was found in two parts of the gradient. First, the majority (about 67%) of the activity was soluble. The second part of the acidic phosphatidate phosphohydrolase was found associated with the protein body fraction (Figure 7). Measurement of other protein body enzymes, such as α-mannosidase or reserve protein, showed that more than 80% of the protein bodies were ruptured during homogenization. Thus, the soluble portion of the pH 5.0 activity could either have originated in the cytosol or be due to leakage from protein bodies.

Because of the doubts in explaining the origin of the soluble acidic phosphatidate phosphohydrolase, Herman and Chrispeels[53] also utilized a different method for the isolation of protein bodies.[54] Broken protoplasts of storage parenchyma cells were layered on discontinuous Ficoll gradients and by this means 70 to 80% of the phosphatidate phosphohydrolase and marker proteins were found to be associated with the protein bodies. Bearing in mind that the total activity of the acidic phosphatidate phosphohydrolase in mung bean cotyledons was considerably more than the pH 7.5 enzyme,[53] then the implication was that the majority of the enzyme was associated with the protein bodies — in common with phospholipase D. In addition, the absence of overlap between some fractions with pH 7.5 or pH 5.0 activity

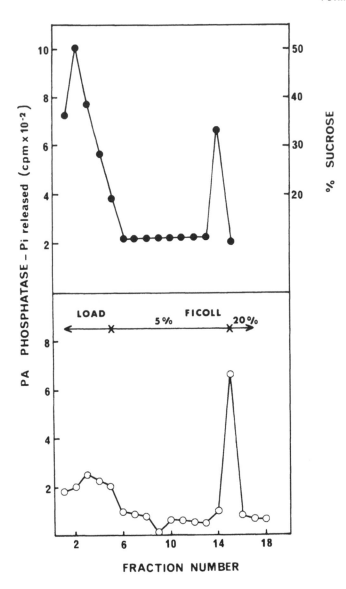

FIGURE 7. Comparison of the distribution of acidic phosphatidate phosphohydrolase activity from mung bean cotyledons on sucrose and Ficoll
density gradients.[53]

demonstrated that two different phosphatidate phosphohydrolases were present in mung bean cotyledons.

Fractions rich in the acidic phosphatidate phosphohydrolase in mung bean also contained acidic *p*-nitrophenolphosphate activity. In order to eliminate the possibility that phosphatidate phosphohydrolysis was due to a nonspecific acid phosphatase, the developmental pattern of the enzyme was examined. Through 6 days of germination, phosphatidate phosphatase increased sixfold in activity in two stages while *p*-nitrophenylphosphatase did not increase for 3 days and then doubled (Figure 8). Together with enzyme purification data, these results indicated that the acidic phosphatidate phosphatase of mung beans was not due to acid phosphatase activity.

Partial purification of the acidic phosphatidate phosphohydrolase by DEAE-cellulose and by Sephadex G-200 gel filtration resulted in further separation of the enzyme from the acidic

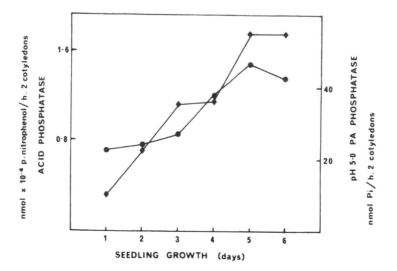

FIGURE 8.   Changes in activity of acidic phosphatidate phosphohydrolase and acid phosphatase during germination of mung bean seedlings. (♦—♦) Phosphatidate phosphohydrolase, (●—●) *p*-nitrophenol phosphate phosphohydrolase. Both enzymes were assayed at pH 5.0.[53]

phosphatase. A broad peak of activity was found by gel filtration of approximately 37,000 mol wt. When assayed in the presence of Nonidet P-40 (octylphenol polyethylene oxide) detergent at pH 5.0, an apparent K_m for phosphatidic acid of 250 μm was found. The pH curve, itself, was broad, peaking at pH 5.0 but still having 60 to 70% of activity at pH 4.0 and 6.0. In spite of the rather high apparent K_m found for the partially purified enzyme, Herman and Chrispeels[53] routinely carried out their phosphatidate phosphohydrolase assays at much lower (2.5 μm) concentrations of phosphatidic acid substrate. It is not known if this could compromise any of their data. They did, however, show that liberation of diacylglycerol from radiolabeled phosphatidylcholine (1 m*M*) at pH 5 (in the presence of detergent), due to the combined action of phospholipase D and phosphatidate phosphohydrolase, was linear for at least 2 hr.

In a similar way to castor bean endosperm[51] and mung bean cotyledons[53] where a pH 7.5 phosphatidate phosphohydrolase was found to be localized in microsomal (endoplasmic reticulum) fractions, a similar localization was reported for optimal generation of phosphatidic acid and diacylglycerol from *sn*-glycerol 3-phosphate in spinach leaf homogenates.[55,56] However, Joyard and Douce[43] showed that purified spinach chloroplasts also synthesize phosphatidic acid and diacylglycerol. Furthermore, this activity occurs almost exclusively in the envelope and is not due to microsomal contamination. These workers purified envelope and thylakoid fractions from intact purified chloroplasts by sucrose density gradient centrifugation. Because the activity of phosphatidate phosphohydrolase was measured indirectly (after incorporation of *sn*-[^{14}C]glycerol 3-phosphate into phosphatidic acid) and because there is a soluble *sn*-glycerol 3-phosphate acyltransferase in chloroplasts, then a reconstituted system had to be used to allow diacylglycerol formation to be measured. This contained envelope membranes plus chloroplast extract to generate the radiolabeled phosphatidate substrate. In addition, inclusion of oleoyl-CoA in the medium caused a shift in the pattern of products formed (Figure 9). Under the latter conditions, phosphatidic acid and diacylglycerol accounted for only 35% of the total radioactivity incorporated whereas in the absence of oleoyl-CoA, these two products accounted for about 80% of the total radioactive products (Figure 9). The accumulation of large amounts of monoacylglycerol in the presence of oleoyl-CoA suggested the presence of a lipid phosphomonoesterase which was active with

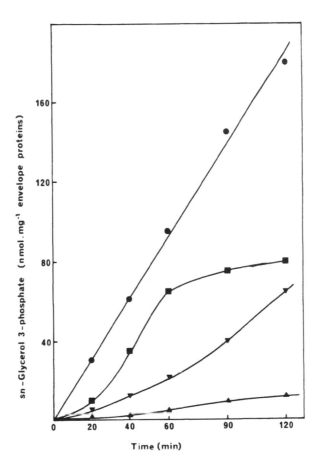

FIGURE 9.   A comparison of the radiolabeling of phosphatidate and diacylglycerol by a mixture of chloroplast envelopes and chloroplast extract in the presence or absence of oleoyl-CoA. (●—●) Phosphatidic acid labeling and (▼—▼) diacylglycerol labeling in the absence of oleoylCoA. (■—■) Phosphatidic acid labeling and (▲—▲) diacylglycerol labeling in the presence of oleoyl-CoA.[43]

a monoacylphosphatidate substrate. It was not clear whether this was separate from phosphatidate phosphohydrolase.

Further work from the same laboratory has provided more information on the spinach chloroplast envelope enzyme.[57] Assays of phosphatidate phosphohydrolase were carried out by first labeling phosphatidate in isolated envelope membranes by the incorporation of radioactivity from sn-[^{14}C]glycerol 3-phosphate. The envelope membranes, loaded with [^{14}C]phosphatidic acid, were then separated from soluble components by centrifugation through a sucrose cushion. The envelopes were resuspended in the appropriate incubation medium and phosphatidate phosphohydrolase activity followed by the disappearance of [^{14}C]phosphatidic acid and/or the appearance of [^{14}C]diacylglycerol.

A pH curve of phosphatidate phosphohydrolase activity showed that activity was detected in the range pH 7.5 to 10.0. Maximal activity was found at pH 9.0 (Figure 10) when the disappearance of phosphatidic acid was associated with the stoichiometric appearance of diacylglycerol (Figure 11). No metal ion requirement was found for the enzyme. In fact, addition of 5 mM MgCl$_2$ caused 75% inhibition and addition of 5 mM EDTA caused a 25% stimulation. The authors drew attention to the lack of a metal ion requirement and the alkaline pH optimum of the chloroplast envelope enzyme which resembled a cytoplasmic membrane

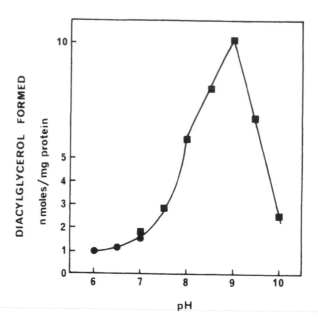

FIGURE 10.   Effect of pH on the activity of phosphatidate phosphohydrolase of spinach envelopes. (●——●) Tris-maleate buffer, (■——■) Tricine-NaOH buffer.[57]

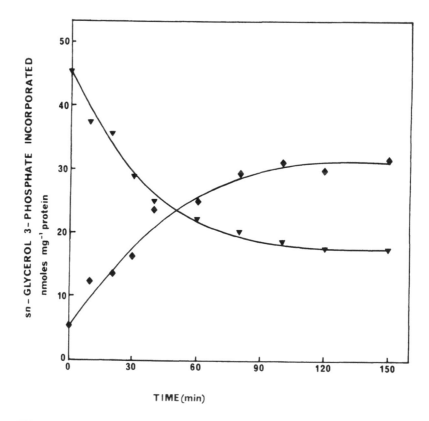

FIGURE 11.   Stoichiometry between phosphatidate loss and diacylglycerol increase during incubation of spinach chloroplast envelopes. (♦——♦) Diacylglycerol, (▼——▼) phosphatidic acid. (From Joyard, J. and Dauce, R., *FEBS Lett.*, 102, 147, 1979. With permission.)

phosphatidate phosphohydrolase in *Bacillus subtilis*.[58] The necessity for prelabeling envelopes as a method of assay for the spinach alkaline phosphatidate phosphohydrolase was caused by the very low activity with exogenous (sonicated) phosphatidic acid.[57]

The localization of spinach chloroplast phosphatidate phosphohydrolase has been more clearly defined. Two membranes (inner and outer envelopes) surround the chloroplast and these can be separated by hyperosmotic treatment of purified chloroplasts, Yeda press treatment, removal of chloroplasts and thylakoids by differential centrifugation, followed by a sucrose density gradient separation of two envelope fractions. The light and heavy fractions are enriched in the outer and inner envelope membranes respectively.[59] Almost no phosphatidate phosphohydrolase was found in the outer membrane fraction whereas the inner membrane fraction showed activities of up to 1.8 nmol/min/mg protein.[59] pH curves were carried out with both membrane fractions and the lack of activity for the outer membrane fraction over a pH range from 6.0 to 9.5 was confirmed. The inner membrane showed a pH optimum of 9.0 as found before[57] and was again inhibited by $MgCl_2$.

IV. THE ROLE OF PHOSPHATIDATE PHOSPHOHYDROLASE IN "16:3" AND "18:3" PLANTS

The division of different plants into "16:3" and "18:3" types was made in Section I in relation to whether or not they contain significant quantities of hexadecatrienoate in their chloroplastic monogalactosyldiacylglycerol. Although plastids always play a very important role in the sequential desaturation of stearate to yield in succession oleate, linoleate, and α-linolenate or of palmitate to yield hexadecatrienoate, certain features of this synthesis, especially in relation to different lipids, are different in the two plant types. Initial results were summarized in Roughan and Slack[13] and only the more recent data are discussed in detail here. It will be seen that the activity of the chloroplast envelope phosphatidate phosphohydrolase is crucial to the overall process.

The essential findings of the early work were that in leaves of 18:3 plants the stearoyl-ACP synthesized in the stroma was desaturated to oleoyl-ACP.[60] The latter thioester was then hydrolyzed,[61] oleate moved across the envelope where it could be activated to a CoA-ester[15] and thus incorporated in phosphatidylcholine in the endoplasmic reticulum. Oleate esterified to phosphatidylcholine could then be desaturated to linoleate and the dilinoleoyl-glycerol moiety incorporated into monogalactosyldiacylglycerol (see Section I). This lipid is a substrate for the third desaturase[19,20] which yields monogalactosyldiacylglycerol with two linolenoyl residues. This galactolipid is characteristic of 18:3 plants such as Asteraceae and Fabaceae.[62] Heinz and Roughan[63] refer to it as eukaryotic monogalactosyldiacylglycerol because the diacylglycerol backbone contains two C18 moieties which originate because of a contribution by the extra-chloroplastic part of the cell.

In 16:3 plants, such as members of the Apiaceae or Brassicaceae,[62] two pathways give rise to monogalactosyldiacylglycerol. First, the eukaryotic pathway described above operates, but it is supplemented to varying degrees by a prokaryotic pathway. The latter is confined to the chloroplast. Acyl chains are incorporated into envelope phosphatidate to result in a 1-oleoyl, 2-palmitoyl molecular species. This is dephosphorylated rapidly to yield diacylglycerol which forms the backbone for monogalactosyldiacylglycerol in which further desaturations take place. Thus, "prokaryotic" monogalactosyldiacylglycerol is typified by palmitate and its desaturation products at C-2[27] and is similar to the lipid in cyanobacteria.[64] This suggests a phylogenetic relationship — hence the term prokaryotic.[63] These two pathways are summarized in Figure 12.

The pathways for 16:3 and 18:3 plants were deduced initially from labeling experiments in leaves.[13,65,66] Additional evidence was provided by experiments using photosynthetically active chloroplasts which retained high rates of fatty acid synthesis.[63] Attempts were made

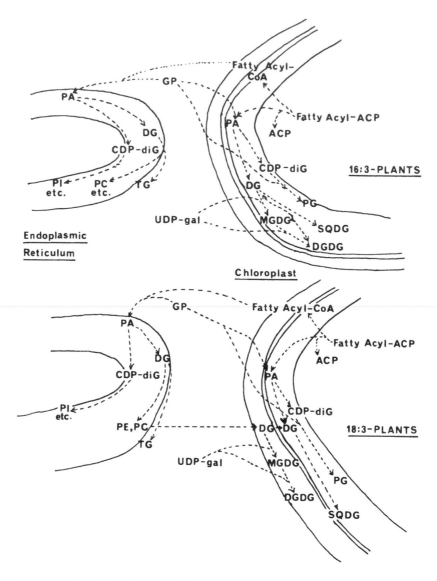

FIGURE 12.   Comparison of glycerolipid synthesis in 16:3 and 18:3 plants. For abbreviations see Figures 2, 3, and 5. The two envelope membranes are shown and the probable location of reactions shown therein.

with a number of species and two 16:3 plants (*Chenopodium album*, and *Solanum nodiflorum*) and two 18:3 plants (*Amaranthus lividus* and *Pisum sativum*) selected because of the high activity of their Percoll gradient-purified chloroplasts. When [1-^{14}C]acetate was used as a precursor and the products of lipid synthesis analyzed, it was noted that the rate of phosphatidic acid synthesis (calculated by combining the radioactivities in phosphatidic acid and diacylglycerol) decreased from 16:3 to 18:3 plants by a factor of two. Even more marked was the difference in phosphatidate phosphohydrolase activity as evidenced by the ratio of radioactive phosphatidate to diacylglycerol. Hydrolysis of phosphatidate decreased by a factor of four on going from 16:3 to 18:3 plants. Indeed, in chloroplasts from 18:3 plants a large proportion of phosphatidic acid was not hydrolyzed even after prolonged (75 min) incubations (Figure 13). The small amounts of diacylglycerol formed in 18:3 plants did not appear to be due to a lack of phosphatidate but was more likely a consequence of low

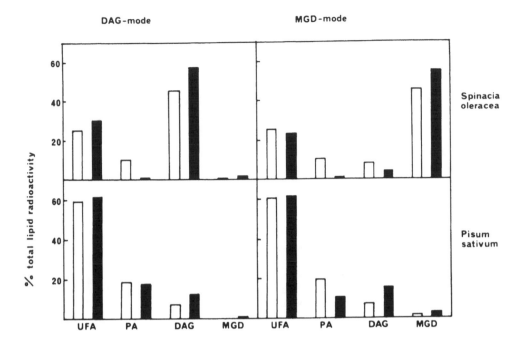

FIGURE 13. Hydrolysis of radiolabeled phosphatidic acid during lipid synthesis in isolated chloroplasts from 16:3 and 18:3 plants. *Spinacia oleracea* (spinach) is a 16:3 plant and *Pisum sativum* (pea) is an 18:3 plant. Open bars = 15 min incubation, solid bars = 30 min incubation. MGD-mode corresponds to incubations in the presence of UDP-galactose. Abbreviations: UFA, unesterified fatty acids; PA, phosphatidic acid; DAG, diacylglycerol; MGD, monogalactosyldiacylglycerol.[63]

phosphatidate phosphohydrolase activity. This conclusion was supported by the observation that measurable phosphatidate phosphohydrolase in pea chloroplasts was low compared to spinach.[45]

Heinz and Roughan[63] also examined the nature of the esterified radiolabeled fatty acids in each lipid class. This was to obtain information about the nature of the substrates for fatty acid desaturation and to see if there were differences in combinations of acids on the glycerol backbone of lipids synthesized by chloroplasts from 16:3 and 18:3 plants. As expected, the fatty acid pattern of diacylglycerol was similar to precursor phosphatidate in all plants. These lipids had increased amounts of palmitate when compared either to the nonesterified fatty acid or monoacylphosphatidate pools. (The latter lipid was enriched in oleate, thus reflecting the specificity of the glycerol 3-phosphate acyltransferase.) No significant differences were seen between 16:3 and 18:3 plants in the fatty acids incorporated into the phosphatidic acid and diacylglycerol pools by isolated chloroplasts.

The lack of differences in fatty acid labeling between the two types of chloroplasts described above contrasts with the nature of the fatty acids accumulated in vivo.[27] By the "eukaryotic" theory of fatty acid synthesis, isolated chloroplasts from 18:3 plants would be unable to carry out adequate desaturation of fatty acids because they lack the complete enzymatic machinery to do so. Such plants could only synthesize linolenate if supplied with exogenous linoleate.[19,20] Confirmation of this was shown in experiments with isolated chloroplasts from 16:3 and 18:3 plants where only the former synthesized significant linolenate during prolonged incubations.[63] In addition, 16:3 chloroplasts possess the full machinery to synthesize the glyceride backbone of monogalactosyldiacylglycerol, including an active phosphatidate phosphohydrolase.

The substrate selectivity of glycerol 3-phosphate acyltransferase and monoacyl glycerol-3-phosphate acyltransferase in representatives of 16:3 (spinach) and 18:3 (pea) plants was

Table 4
**DISTRIBUTION OF RADIOACTIVITY FROM
[1-^{14}C]ACETATE IN LIPID CLASSES OF
CHLOROPLASTS FROM 16:3 AND 18:3 PLANTS**

	Distribution of radioactivity (%)		
Plant	Fatty acids	Phosphatidate	Diacylglycerol
16:3-type			
Spinacia oleracea	26	23	31
Solanum nodiflorum	23	15	47
Transitional 16:3			
Chenopodium quinoa	57	17	6
18:3-type			
Carthamus tinctorius	58	22	2
Pisum sativum	62	16	3

Note: Isolated chloroplasts were incubated in the light for 5 min. Minor products
are omitted from the table. The data are taken from Gardiner and Roughan[68]
where full details can be found.

studied by Frentzen et al.[45] In agreement with the labeling studies referred to above, no differences were seen between chloroplasts from these species. In both pea and spinach chloroplasts, glycerol 3-phosphate acyltransferase esterified the *sn*-1 position and preferred oleate as substrate with oleoyl-ACP being used about ten times as rapidly as oleoyl-CoA. Monoacyl glycerol 3-phosphate acyltransferase again preferred ACP-esters but transferred palmitate to the *sn*-2 position. These results have been confirmed by labeling studies with isolated chloroplasts from a number of 16:3 and 18:3 plants.[67] The result of these activities was that phosphatidic acid in both 16:3 and 18:3 plants contained mainly 1-oleoyl, 2-palmitoyl combinations. The authors also noted the low conversion of phosphatidic acid to monogalactosyldiacylglycerol in the presence of UDP-galactose in pea chloroplasts and concluded that this reflected the low phosphatidate phosphohydrolase activity which they measured at pH 9.0.[45]

The relationship of chloroplast phosphatidate phosphohydrolase to lipid synthesis in 16:3 and 18:3 plants was examined further.[68] Chloroplasts were isolated from five species which were categorized as 16:3 plants (*Spinacia oleracea* and *Solanum nodiflorum*), a transitional 16:3 plant (*Chenopodium quinoa*), and 18:3 plants (*Carthamus tinctorius* and *Pisum sativum*). Organelles were incubated in the light for 5 min with [1-^{14}C]acetate in order to label phosphatidate and then transferred to darkness when the loss of radioactivity was followed. The time and conditions of the light incubations were chosen to give optimal labeling of phosphatidate. These experiments resulted in 15 to 23% of the total radioactivity incorporated into lipids being present in phosphatidate. In contrast to the similar level of incorporation into phosphatidate, only the 16:3 plants showed high amounts of radiolabel in diacylglycerol (Table 4) as predicted if they contained high amounts of phosphatidate phosphohydrolase. When the chloroplasts were transferred to the dark, fatty acid synthesis stopped,[31] but a redistribution of label between lipid classes occurred. In particular the radioactivity in phosphatidate declined and that in diacylglycerol increased. The changes in radioactivity in these two classes did not exactly match, but it was concluded that there was no evidence for any breakdown of phosphatidic acid by any route other than by phosphatidate phosphohydrolase. A slow deacylation of diacylglycerol had been observed before[46] and this was thought likely to account for the difference. Therefore, the loss in counts from phosphatidate was taken to indicate phosphatidate phosphohydrolase activity. The changes in radioactivity associated

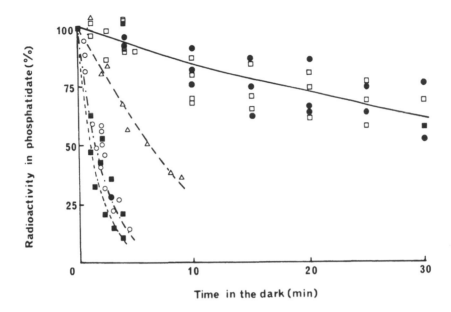

FIGURE 14. Phosphatidate decay in darkened chloroplasts. Illuminated chloroplasts were preincubated with [14C]acetate in the light for 5 min and then incubated in the dark for the times indicated. (●——●) *Pisum sativum* (pea) and (□——□) *Carthamus tinctorius* (safflower) are 18:3 plants. (△——△) *Chenopodium quinoa* (Quinoa) is an intermediate plant. (○——○) *Spinacia oleracea* (spinach) and (■——■) *Solanum nodiflorum* are 16:3 plants. (From Gardiner, S. E. and Roughan, P. G., *Biochem. J.*, 210, 949, 1983. With permission.)

with phosphatidate in the five different chloroplasts are shown in Figure 14. It will be seen that radioactivity declines quickly in the two 16:3 plants, *Solanum nodiflorum* and *Spinacia oleracea*, at an intermediate rate in the transitional 16:3 plant, *Chenopodium quinoa*, and slowly in the two 18:3 plants, *Pisum sativum* and *Carthamus tinctorius*. Because the reduction in radioactivity appeared to be first order, a linear regression analysis was performed in order to obtain rate constants and half life values for phosphatidate. These data showed that half lives for phosphatidate in *Spinacea oleracea* and *Solanum nodiflorum* were only 1.3 and 1.7 min respectively, whereas in the two 18:3 plants half lives of 39 min were found (Table 5). *Chenopodium quinoa* had phosphatidate with an intermediate half life (6 min) which corresponded to the small but significant quantities of hexadecatrienoate in its monogalactosyldiacylglycerol (Table 5). Similar results were also obtained when incubations were carried out entirely in the light but with a pulse-chase of acetate after 5 min to prevent any further incorporation of radioactivity into lipids during the subsequent incubation. Thus, the phosphatidate decay in the dark was similar to that in the light and values obtained for the half lives were not influenced by the incubation conditions.[68]

The low capacity of chloroplasts isolated from 18:3 plants to accumulate diacylglycerol, therefore, was a direct reflection that phosphatidate phosphohydrolase was poorly active. Thus, half lives for chloroplastic phosphatidate were up to 30-fold shorter in 16:3 plants which provided, therefore, substrate diacylglycerol for galactolipid synthesis. This, in turn, allowed desaturation to polyunsaturated fatty acids to occur at both *sn*-1 and *sn*-2 positions of the glycerol. Since phosphatidate which gave rise to the diacylglycerol backbone contained palmitate at the *sn*-2 position, then sequential desaturation of this acid yielded hexadecatrienoate. In contrast, 18:3 plants which use an extra-chloroplastic route for some of their desaturation (Figure 12) provide the chloroplast in vivo with a diacylglycerol for galactolipid synthesis which does not contain any C16 fatty acid at the *sn*-2 position.

Confirmation of the above results and conclusions has been obtained for different 16:3

Table 5
PHOSPHATIDATE DEPHOSPHORYLATION IN 16:3
AND 18:3 PLANTS[68]

Plant	$T_{1/2}$ phosphatidate (min)	1-18:3, 2-16:3-MGDG (% total MGDG)
16:3 plants		
Solanum nodiflorum	1.3	40
Spinacea oleracea	1.7	50
Transitional 16:3		
Chenopodium quinoa	5.8	8
18:3 plants		
Pisum sativum	38.5	0
Carthamus tinctorius	38.5	1

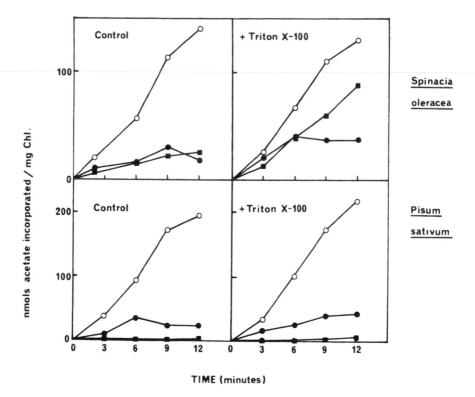

FIGURE 15.   Effect of Triton X-100 on the time course of incorporation of radioactivity from [1-^{14}C]acetate into lipids of chloroplasts isolated from spinach or pea. (O—O) Unesterified fatty acids, (●—●) polar lipids, (▲—▲) diacylglycerol. Incubations were carried out in light ± 0.13 mM Triton X-100. Data taken from Reference 70 where further details can be found.

and 18:3 plants[69] and by manipulating the proportions of lipid products produced from [1-^{14}C]acetate by isolated chloroplasts from 16:3 and 18:3 plants.[70] Chloroplasts from 18:3 plants incorporated more radioactivity into nonesterified fatty acids and less into diacylglycerol than those from 16:3 plants in all cases (Table 4). When exogenous glycerol 3-phosphate (0.25 mM) or Triton X-100 (0.13 mM) were added, increased glycerolipid synthesis was seen in all chloroplasts. However, chloroplasts from 18:3 plants accumulated radioactivity in phosphatidate rather than in diacylglycerol as was found for chloroplasts from 16:3 plants (Figure 15). Thus, the inability of chloroplasts from 18:3 plants to dephosphorylate phosphatidate is a major cause of the inability to synthesize "prokaryotic" monogalactosyldiacylglycerol. In addition, it has also been found that such chloroplasts are relatively poor at incorporating their small amounts of diacylglycerol into monogalactosyldiacylglycerol when compared to chloroplasts from 16:3 plants.[70] An explanation of this could lie in the fact that whereas both phosphatidate phosphohydrolase and UDP-galactose:diacylglycerol galactosyltransferase are found in the inner envelope membrane of spinach[8] the two enzymes may be present in different membranes in 18:3 plants. Thus, the UDP-galactose:diacylglycerol galactosyltransferase of pea chloroplasts has been localized on the outer envelope membrane.[71] This location would also cause 18:3 chloroplasts to use imported diacylglycerol for galactolipid synthesis rather than substrate generated by chloroplastic phosphatidate phosphohydrolase.

The low activity of the chloroplastic phosphatidate phosphohydrolase in 18:3 plants does not seem to be mimicked by either a low activity of the acyl transferases forming phosphatidate[45,70] nor in a low activity of the CTP:phosphatidate cytidylyltransferase.[25] However, there may be differences in, for example, phosphatidylglycerol synthesis in chloroplasts from 16:3 (spinach) and 18:3 (pea) plants due to the interaction of the various pathways. For example, the low ability of spinach envelope membranes to make phosphatidylglycerol[25] may be due to competing phosphatidate phosphohydrolase and the ability of Mn^{2+} + Mg^{2+} to stimulate phosphatidylglycerol synthesis in spinach[24] but not pea; this may again reflect the active phosphatidate phosphohydrolase in spinach which can be inhibited by $MgCl_2$[57] and may well be sensitive to Mn^{2+} also.

V. THE ROLE OF PLANT PHOSPHATIDATE PHOSPHOHYDROLASE IN SEED OIL ACCUMULATION

Seed oils have great industrial importance. Although the majority of such fats contain palmitate, oleate, and linoleate as their principal fatty acids, certain unusual acyl groups are major constituents in some seeds.[6,72] The only economically important seed oil which is not triacylglycerol is that from jojoba (*Simmondsia chinensis*) where it is a wax ester.

Early experiments by Barron and Stumpf[56] confirmed the existence of the Kennedy pathway for triacylglycerol synthesis in avocado mesocarp fractions. This has been confirmed for other seeds and establishes the importance of phosphatidate phosphohydrolase in seed oil formation. Studies of the enzyme's activity indirectly through radiolabeling experiments are, however, particularly difficult because of a rapid exchange of diacylglycerol with the diacylglycerol backbone of phosphatidylcholine. The most likely cause of this is the reversible action of cholinephosphotransferase.[73] In addition, the esterification of monoacylphosphatidylcholine by acyl-CoAs has been postulated to be important either in synthesis of phosphatidylcholine[74,75] or for exchange of acyl residues at the *sn*-2 position.[76] Because of the accepted importance of phosphatidylcholine as a substrate for oleate desaturation and since many seeds accumulate oils rich in linoleate (e.g., 55% maize, 60% soybean, 70% sunflower) then the equilibration of the diacylglycerol pool with phosphatidylcholine has considerable importance, rather than diacylglycerol merely being a product of phosphatidate phosphohydrolase en route for triacylglycerol. Although some workers have found evidence against such an equilibration,[77] it has been accepted generally for oil seeds.[13,78] Phosphatidylcholine has also been proposed as a substrate for linoleate desaturation in oil seeds[79] which contrasts to the role of monogalactosyldiacylglycerol for this desaturation in leaf tissues (Figure 3). However, linolenate is a minor component of most commercially important seed oils[80] where its presence is regarded as undesirable.

The specific incorporation of different fatty acids into various positions of triacylglycerol takes place by pathways which are not entirely clear at present. Different laboratories have their own opinions as to the exact details of the processes. The ideas of two of the main

(a)

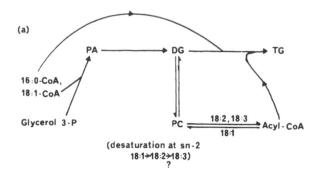

(b)

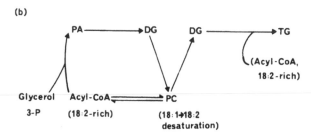

FIGURE 16. Synthesis of highly unsaturated triacylglycerols — comparison of two proposed mechanisms. In scheme (a) phosphatidic acid enriched in palmitate and oleate is dephosphorylated to diacylglycerol. The latter equilibrates with phosphatidylcholine where desaturation of oleate (and, maybe, linoleate) at the *sn*-2 position can take place. Therefore, the diacylglycerol pool can also become more unsaturated. Transfer of oleate from oleoyl-CoA to phosphatidylcholine also provides substrate for desaturation and linoleate (and linolenate) can be returned to an acyl-CoA pool to participate (with other acyl-CoAs) in diacylglycerol acylation.[13] In scheme (b) phosphatidylcholine is again involved in desaturation of oleate to linoleate. Unsaturated diacylglycerols are generated by the back reaction of cholinephosphotransferase. Linoleate is preferentially transferred to the acyl-CoA pool which gradually becomes enriched in this acid and so results in a linoleate-enriched phosphatidic acid pool. The operation of the cycle (phosphatidate phosphohydrolase-cholinephosphotransferase-lyso PC acyltransferase-glycerol 3-P acyltransferase-lyso PA acyltransferase) therefore causes a gradual enrichment of linoleate in the diacylglycerol pool used for triacylglycerol synthesis.[78]

laboratories concerned are summarized in Figure 16. Roughan and Slack[13] suggest that the equilibration of the diacylglycerol pool with phosphatidylcholine via the activity of cholinephosphotransferase[73] is the prime mechanism for polyunsaturated fatty acid enrichment of seed oils. On the other hand Stymne and co-workers[81] proposed that linoleate formed by desaturation of oleate at the *sn*-2 position of phosphatidylcholine was returned to the acyl-CoA pool by acyl exchange. This linoleoyl-CoA was then selectively utilized for phosphatidic acid synthesis and was returned to the diacylglycerol pool (and either triacylglycerol or phosphatidylcholine) by the action of phosphatidic acid phosphohydrolase.

The general applicability of the Kennedy pathway of triacylglycerol synthesis in seeds[82] with possible modification as mentioned above has been accepted generally. The initial acylation of glycerol 3-phosphate was localized in the endoplasmic reticulum of castor bean seeds.[83] (This membrane is also the site of acyl exchange reactions, cholinephosphotransferase, and oleate desaturation.) In assays of the castor bean glycerol 3-phosphate transferase using palmitoyl-CoA, no monoacyl glycerol 3-phosphate was detected and it was proposed that the enzyme could acylate both the *sn*-1 and *sn*-2 positions by a concerted mechanism.[83] However, it has been pointed out that palmitate is essentially absent from the *sn*-2 position of mature seed oil triacylglycerols[6] and from diacylglycerol in developing cotyledons. In order to account for this positional specificity, two separate acyltransferases acting sequentially would be necessary[13] with the second enzyme selecting for C18 fatty acids. Such enzymes have been reported recently from spinach leaves.[84] The glycerol 3-phosphate acyltransferase was solubilized and partly purified from endoplasmic reticulum and the monoacyl glycerol 3-phosphate acyltransferase was studied using membrane fractions. In contrast, to the chloroplast enzymes (Section IV) both endoplasmic reticulum enzymes used acyl-CoAs. The glycerol 3-phosphate acyltransferase transferred both palmitoyl and oleoyl groups to the *sn*-1 position. Although palmitate was the preferred substrate, the in vivo preponderance of oleate in the extra-chloroplastic compartment guarantees that the *sn*-1 position of phosphatidic acid will contain oleate in addition to palmitate. The second acyltransferase also used acyl-CoA substrates but while oleoyl-CoA and linoleoyl-CoA gave similar activities, the rates with palmitoyl-CoA were noticeably lower.[85] These results show the essential differences between the plastid (Section IV) and endoplasmic reticulum acyltransferases and provide a rationale for the existence of phosphatidic acid containing palmitate essentially confined to the *sn*-1 position. The monoacyl glycerol 3-phosphate acyltransferase has also the right specificity to return linoleoyl-CoA to the phosphatidic acid pool as suggested by Stymne et al.[81] (Figure 16).

The dephosphorylation of phosphatidic acid during triacylglycerol formation by microsomal fractions from safflower seeds was studied by Stymne et al.[81] for three different varieties. These were a high oleate (75% oleate, 16% linoleate), Gila (16% oleate, 75% linoleate), and a very high linoleate variety (6% oleate, 88% linoleate). A typical time-course experiment is shown in Figure 17 where microsomes were capable of acylating added glycerol 3-phosphate in the presence of [14C]linoleoyl-CoA and the radiolabeled phosphatidic acid was rapidly dephosphorylated to yield diacylglycerol which was itself then acylated to give triacylglycerol. An examination of the ability of different acyl-CoAs to be incorporated into phospholipids and acylglycerols in Gila and high oleate varieties also gave some information about the substrate specificities of the enzymes involved — including phosphatidate phosphohydrolase. Results from such an experiment are shown in Table 6. Examination of the data reveals that while acylation of glycerol 3-phosphate was approximately equally high with each unsaturated acyl-CoA, when a mixture of acyl-CoAs was presented linoleoyl-CoA was preferred. Incorporation of radioactivity into phosphatidic acid and diacylglycerol showed a similar ratio of radioactivity incorporated from each single acyl-CoA substrate. This suggested strongly that the microsomal phosphatidate phosphohydrolase did not show a marked substrate specificity. The fatty acid labeling patterns of phosphatidic acid and diacylglycerol (and, because of cholinephosphotransferase reversibility, phosphatidylcholine) therefore, were very similar. The diacylglycerol acyltransferase, which was suggested earlier[13] to have no special substrate specificity, incorporated all acyl-CoAs so that the final triacylglycerol acyl composition was different from that of the precursor phosphatidic acid and diacylglycerol (Table 6).

Further experiments with safflower and also with developing cotyledons of sunflower have confirmed the overall conclusions made above.[85,86] However, in avocado microsomes which synthesize triacylglycerol by the Kennedy pathway, an absence of acyl exchange and diacylglycerol-phosphatidylcholine interconversion was found.[85] The overall impact of the various enzymes on triacylglycerol fatty acyl quality is summarized in Figure 18. Specificity in acyl chain enrichment appears to arise from the glycerol 3-phosphate acyltransferase and monoacylglycerol 3-phosphate acyltransferases coupled with desaturation of oleate in phos-

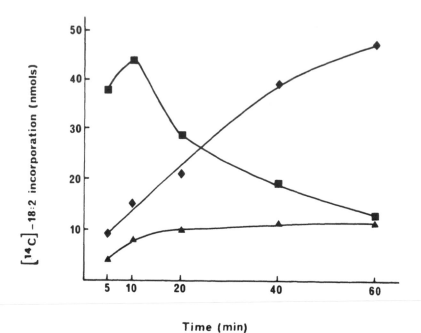

Time (min)

FIGURE 17.   Time course of labeling of phosphatidic acid and acylglycerols of safflower microsomes with [^{14}C]linoleoyl-CoA. (■——■) Phosphatidic acid, (▲——▲) diacylglycerol, (◆ — ◆) triacylglycerol. Incubations were carried out at 30°C and pH 7.2 in the presence of 0.4 m*M* glycerol 3-phosphate. Albumin was present to protect the substrate from hydrolysis by a microsomal thioester hydrolase.[81]

Table 6
LABELING OF PHOSPHOLIPIDS AND ACYLGLYCEROLS FROM DIFFERENT ACYL-CoAS BY SAFFLOWER MICROSOMES

	^{14}C-labeled fatty acids (nmol)							
	PC		PA		DG		TG	
Acyl-CoA substrate	Gila	High (18:1)	Gila	High (18:1)	Gila	High (18:1)	Gila	High (18:1)
Stearoyl-CoA	18.5	10.4	11.9	6.6	1.1	1.2	10.1	9.0
Oleoyl-CoA	62.5	55.5	56.4	33.9	5.1	5.2	12.8	14.2
Linoleoyl-CoA	55.3	44.2	61.0	53.5	6.4	6.2	10.7	11.5
Linolenoyl-CoA	61.2	34.7	39.0	49.0	4.8	4.6	13.2	14.9
Mixture								
Stearoyl-CoA	0	0	2.3	3.8	—	—	—	
Oleoyl-CoA	31.2	22.3	8.9	7.9	—	—	—	
Linoleoyl-CoA	11.9	10.4	16.4	17.7	—	—	—	
Linolenoyl	9.7	11.7	9.1	11.2	—	—	—	

Note: Microsomes were prepared from safflower cotyledons 14 to 18 days after flowering and incubated in phosphate buffer at pH 7.2 with 250 nmol of [^{14}C]acyl-CoA as single substrate or 60 nmol each for mixed substrate. Glycerol 3-phosphate was present at 400 nmol. Abbreviations: PC, phosphatidylcholine; PA, phosphatidic acid; DG, diacylglycerol; TG, triacylglycerol. Data for two safflower varieties are shown. Gila is a normal variety accumulating triacylglycerols which contain 16% 18:1 and 75% 18:2. The high (18:1) variety accumulates triacylglycerols with 75% 18:1 and 16% 18:2.

From Stymne, S., et al., *Biochim. Biophys. Acta,* 752, 198, 1983. With permission.

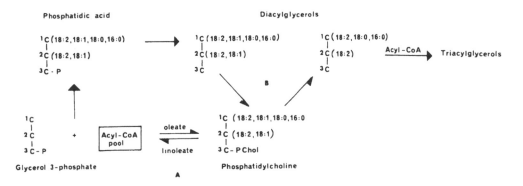

FIGURE 18.   Regulation of acyl quality of triacylglycerol in developing safflower cotyledons. (A) Acyl exchange which controls the movements of oleate and linoleate between the *sn*-2 position of phosphatidylcholine and the acyl-CoA pool. (B) The equilibration of diacylglycerol and phosphatidylcholine is presumed to be catalyzed by choline phosphotransferase. Because of the desaturation of oleate when it is esterified to the *sn*-2 position of phosphatidylcholine, the diacylglycerol returned to the diacylglycerol pool is enriched in linoleate at that position. (From Stobart, A. K. and Stymne, S., *Planta*, 163, 119, 1985. With permission.)

phatidylcholine and return of resultant linoleate to the acyl-CoA pool. Phosphatidate phosphohydrolase is envisaged as dephosphorylating all available phosphatidic acid species at comparable rates.[78]

Further experiments with the developing safflower have provided more information about phosphatidic acid metabolism in connection with triacylglycerol accumulation. The acylation of glycerol 3-phosphate by [14C]oleoyl-CoA was followed in the microsomal fraction. Incubation with EDTA (10 mM) and assay at pH 7.2 resulted in about 30% more radioactive glycerol in phosphatidate than in controls. At the same time radioactivity accumulated in diacylglycerol (and phosphatidylcholine) was decreased. Incubations with 20 mM MgCl$_2$ produced the opposite effects with decreased radiolabel in phosphatidate and increased labeling of diacylglycerol. Addition of MgCl$_2$ after a period of incubation with EDTA also stimulated phosphatidate dephosphorylation[87] (Figure 19). These results emphasize a difference with the chloroplast envelope (alkaline) phosphatidate phosphohydrolase which is inhibited by Mg^{2+} (see Reference 57 and Section III).

If, as the results have implied, triacylglycerols are produced by operation of a modified Kennedy pathway involving equilibration of phosphatidylcholine with diacylglycerol and, if microsomal phosphatidate phosphohydrolase is nonselective with regard to its phosphatidate substrates, then two features of lipid metabolism can be predicted. First, the substrate specificity of the glycerol 3-phosphate and monoacylglycerol 3-phosphate acylases should regulate the fatty acid composition of endogenous phosphatidate. Second, comparison of the fatty acid compositions of phosphatidate and phosphatidylcholine should reveal the importance of cholinephosphotransferase-induced exchange with diacylglycerol and acylation of monoacylphosphatidylcholine. The endogenous fatty acid compositions of phosphatidate and phosphatidylcholine were found to be identical (Table 7). Furthermore, the relatively small amounts of oleate in these phosphoglycerides point to its preferential use in acylating monoacylphosphatidylcholine and its rapid desaturation in phosphatidylcholine. The similar fatty acid compositions suggest that linoleate when formed in phosphatidylcholine is rapidly returned to the acyl-CoA pool and then appears in phosphatidate; otherwise an enrichment of linoleate in phosphatidylcholine would be seen.

The importance of Mg^{2+} in overcoming the block in phosphatidate utilization caused by EDTA is shown in more detail in Table 8. In the absence of Mg^{2+} virtually no changes were observed in the quantity of diacylglycerol and phosphatidate labeled during a 2-hr incubation. On the other hand, addition of Mg^{2+} caused an increase in diacylglycerol quantity

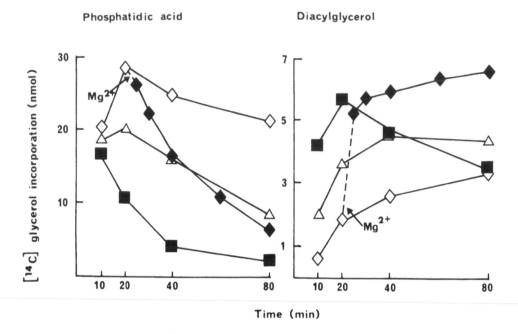

FIGURE 19.   Effect of EDTA and Mg^{2+} on phosphatidate dephosphorylation in microsomes from developing safflower seeds. Microsomes incubated with [^{14}C]glycerol 3-phosphate in the presence of (◊—◊) 10 mM EDTA and (■—■) 20 mM $MgCl_2$. Microsomes incubated with substrates and EDTA for 20 min before addition of (◆—◆) 20 mM $MgCl_2$ or (△—△) No addition (control sample). Taken from Reference 87 where further details can be found.

Table 7
COMPARISON OF THE FATTY ACID
COMPOSITIONS OF PHOSPHATIDYLCHOLINE
AND PHOSPHATIDIC ACID IN MICROSOMES
FROM DEVELOPING SAFFLOWER SEEDS[87]

	Distribution (% total fatty acids)			
	16:0	18:0	18:1	18:2
Phosphatidylcholine	12 ± 0.3	3 ± 0.2	1 ± 0.1	84 ± 0.4
Phosphatidic acid	13 ± 0.6	2 ± 0.1	2 ± 0.1	83 ± 0.7

which was equivalent to the decline in phosphatidate amounts.[88] Although interpretation of these experiments is complicated by the net flow of carbon through diacylglycerol to triacylglycerol, they emphasize the requirement for Mg^{2+} of microsomal phosphatidate phosphohydrolase.

To summarize, oil production in seeds takes place by the Kennedy pathway with phosphatidate phosphohydrolase playing a key role. However, the phosphohydrolase does not influence the fatty acid composition of the accumulated triacylglycerol. The microsomal phosphatidate phosphohydrolase responsible for oil synthesis also seems to have distinctly different properties from the plastid enzyme.

VI. PHOSPHATIDATE PHOSPHOHYDROLASE IN MICROORGANISMS

A. Bacteria

Phosphatidic acid is an essential intermediate in the biosynthesis of glycerophosphatides

Table 8
EFFECT OF Mg²⁺ ON THE
DEPHOSPHORYLATION OF
PHOSPHATIDIC ACID IN EDTA-
TREATED MICROSOMAL
PREPARATIONS OF DEVELOPING
SAFFLOWER COTYLEDONS

Pretreatment	PA (nmol)	Diacylglycerol (nmol)
No oleoyl-CoA		
Zero time	2	2
120 min + Mg²⁺	4	4
+ oleoyl-CoA		
Zero time	22	8
30 min + Mg²⁺	8	20
35 min − Mg²⁺	21	8
120 min + Mg²⁺	7	23
125 min − Mg²⁺	27	9

Note: Microsomes were pretreated with EDTA, *sn*-glycerol 3-phosphate (and oleoyl-CoA) and then incubated with monoacylphosphatidylcholine and [¹⁴C]oleoyl-CoA for 5 min. Further incubations were then as shown with or without 20 mM Mg²⁺.

From Stobart, A. K. and Stymne, S., *Biochem. J.*, 232, 217, 1985. With permission.

in *Escherichia coli* and labeling studies with [³H]glycerol have shown that the phosphatidic acid pool was both rapidly labeled and turned over. Phosphatidic acid was then used to produce CDP-diacylglycerol but there was no evidence of a diacylglycerol pool produced by the action of phosphatidic acid phosphohydrolase.[89]

Later work with *E. coli* has shown the presence of a monoacylphosphatidic acid phosphohydrolase which also has some activity towards phosphatidic acid.[90] *sn*-Glycerol 3-phosphate was acylated by palmitoyl-ACP to produce the monoacylphosphatidic acid, which was then either further acylated to give phosphatidic acid or dephosphorylated to give monoacylglycerol, depending on the experimental conditions and the choice of acyl donor. Optimal dephosphorylation of monoacylphosphatidic acid occurred at pH 7.0. The monoacylphosphatidic acid phosphohydrolase differed from *E. coli* alkaline phosphatase in that it was particulate and was not inhibited by inorganic phosphate. The dephosphorylation activity towards monoacylphosphatidate and phosphatidate did not require divalent cations and was not affected by 5 mM NaF, but was dependent on sulfhydryl groups for activity. No conversion of monoacylphosphatidylethanolamine or phosphatidylethanolamine to monoacylglycerol or diacylglycerol occurred which suggests that phospholipase C was not responsible for the dephosphorylation activity.

Blank and Snyder[49] have reported an alkaline phosphatase in *E. coli* which had activity towards monoacylphosphatidic acid and other lipid substrates with monoester phosphate moieties, but which were not acylated at the *sn*-2 position. It had no activity towards phosphatidic acid (Table 9). The differences in assay conditions for the monoacylphosphatidic acid phosphatase and alkaline phosphatase used by the two groups of workers do not explain why the alkaline phosphatase had no activity towards phosphatidate.

A phosphatase enzyme with activity towards phosphatidate at pH 8.0 has been described

Table 9
COMPARISON OF SUBSTRATE SPECIFICITIES OF ALKALINE
PHOSPHATASE FROM *E. COLI*[49] AND PHOSPHATIDATE
PHOSPHATASE FROM *S. CEREVISIAE*[96]

Substrate	Concentration (mM)	Activity	
		E. coli alkaline phosphatase[a]	*S. cerevisiae* phosphatidate phosphatase[b]
1-Alkyl-2-acylglycerol-3-P	5	1.2	—
1-Acylglycerol-3-P	0.33	87	—
	0.2	—	5
	0.4	—	15
1-Alkylglycerol-3-P	5	63	—
1-Alkyldihydroxyacetone-P	10	49	—
Phosphatidylcholine	0.43	0	—
Phosphatidate	0.22	0	—
	0.2	—	83
	0.4	—	100
Phosphatidylinositol 4-P	0.2	—	0
	0.4	—	0
Phosphatidylinositol 4,5-bis P	0.2	—	0
	0.4	—	0
rac-Glycerol 1-P	0.2—1.0	—	0
Glycerol 2-P	0.2—1.0	—	0
5'-AMP	0.2—1.0	—	0
3'-AMP	0.2—1.0	—	0
Phosphoserine	0.2—1.0	—	0
Glucose 6-P	0.2—1.0	—	0
p-Nitrophenylphosphate	0.2—1.0	—	0

[a]   Percent dephosphorylated in 1 hr.
[b]   Percent values obtained with 0.4 mM phosphatidate as substrate.

in the cytoplasmic membranes of *Bacillus subtilis*.[58] It was similar to the monoacylphos-phatidate phosphohydrolase from *E. coli* in that it had no requirement for divalent metal ions (in fact 1 mM Mg^{2+} caused 65% inhibition) but there was no report of its substrate specificity. It had a biphasic saturation curve for phosphatidate at 0.12 mM and 0.4 to 0.5 mM, but in the presence of another lipid or a high ionic strength, saturation only occurred at the higher level. Its optimal pH was 8 to 8.6 and it was more active in Tris-HCl or Tris-maleate than in acetate or glycine buffers. The presence of the nonionic detergent, Cutscum inhibited the enzyme by 50% at concentrations of 0.03 and 0.2%, but stimulated activity tenfold at 0.125%. Lipids were added to the reaction at lipid to substrate ratios of 1:1 and 5:1, diglucosyldiacylglycerol and phosphatidylglycerol gave no inhibition at 1:1, but 40 to 60% inhibition at 5:1. Diacylglycerol and phosphatidylethanolamine gave 67 and 90% in-hibition at 1:1 and 5:1, respectively.

The role of a phosphatidate phosphohydrolase in a Gram positive organism, such as *B. subtilis*, is clearer because *B. subtilis* contains neutral lipids (including diacylglycerol) and glycosyldiacylglycerols which require diacylglycerol as a precursor. The diacylglycerol is produced from phosphatidate, the sole lipid precursor, by the action of the phosphohydrolase. However, in Gram negative organisms, such as *E. coli*, where phosphoglyceride synthesis is via CDP-diacylglycerol and a metabolically active pool of diacylglycerol has not been demonstrated, the role of a specific phosphatidate phosphohydrolase is not obvious.

B. Fungi and Yeasts

Phosphatidate had been presumed to be the precursor for diacylglycerol biosynthesis in fungi and yeasts[91] and the presence of a phosphatidate phosphohydrolase has been demonstrated in two pathogenic dermatophytic fungi, *Microsporum gypseum* and *Epidemophyton floccosum*.[92] In both fungi phosphatidate phosphohydrolase activity was found in mitochondrial and microsomal fractions, with higher activity being present in the microsomes. Microsomal activity in the two fungi had a broad pH optimum of pH 5.0 to 7.0, while the mitochondrial activity had a pH optimum of pH 6.0. The enzyme from *E. floccosum* had apparent K_m's of 0.26 and 0.25 mM for mitochondrial and microsomal activity respectively, and the enzyme from *M. gypseum* had a mitochondrial K_m of 2.20 mM and a microsomal K_m of 1.42 mM for phosphatidate.

Phosphatidate phosphohydrolase activity in both fungi, except for the mitochondrial activity of *E. floccosum*, was stimulated by Mg^{2+}. The presence of other divalent cations, such as Fe^{2+}, Hg^{2+}, Ba^{2+}, Cu^{2+}, and Mn^{2+} had inhibitory effects, whereas the phosphatidate phosphohydrolase from neither fungi was susceptible to sulfhydryl reagents. Phosphatidylcholine, monoacylphosphatidylcholine, and phosphorylcholine had inhibitory effects but glycerol 3-phosphate did not.

In studies on the synthesis of phosphatidic acid in bakers yeast, Kuhn and Lynen[93] reported no loss of phosphatidic acid due to phosphohydrolase activity and postulated that in yeast phosphatidate phosphohydrolase only had low activity towards phosphatidate containing saturated acyl groups.

A comprehensive in vitro study of lipids labeled from sn-[^{14}C]glycero-3-phosphoric acid by Steiner and Lester[94] concluded that there was indirect evidence for the production of diacylglycerol from phosphatidic acid in *Saccharomyces cerevisiae*. When unlabeled CDP-choline was added to the system, the incorporation of sn-[^{14}C]glycero-3-phosphoric acid into phosphatidylcholine increased by a factor of 3.9. However, under the same conditions no sn-glycero-3-[^{32}P]phosphoric acid was incorporated into phosphatidylcholine, which is consistent with the production of phosphatidylcholine from CDP-choline and diacylglycerol. The diacylglycerol arose from phosphatidic acid as a result of phosphatidate phosphohydrolase activity.

A further phospholipid labeling study using *S. cerevisiae* mentions the presence of phosphatase activity.[95] However, although the phosphatase had activity towards phosphatidate at pH 7.2, it also had activity towards sn-[2-^{3}H]glycerol 3-phosphate, and was not, therefore, a specific phosphatidate phosphohydrolase.

Recent work by Hosaka and Yamashita[96] has demonstrated the presence of a specific phosphatidate phosphohydrolase in *S. cerevisiae*. The enzyme was present in both soluble (specific activity 2 to 3 mU/mg protein) and membrane fractions (specific activity 4 to 6 mU/mg protein) and has been purified 600-fold from the soluble fraction. The enzyme in the membrane fraction could be solubilized using cholate, deoxycholate, and Triton X-100, but not by sonication or treatment at pH 9.0 to 9.5. The enzyme present in both fractions was stable for at least one month at $-80°C$.

Partial purification of the soluble enzyme yielded two fractions from Blue Sepharose chromatography. The two fractions were similar in terms of metal ion requirements, substrate specificity, and reaction products. The presence of activity in the two fractions may be due to isozymes or to the purification method used.

The partially purified phosphatidate phosphohydrolase had a pH optimum of 7.5 and an apparent K_m for phosphatidate of 0.05 mM. The presence of Mg^{2+} in the assay was an almost absolute requirement and Mn^{2+} was slightly stimulatory, but Ca^{2+}, Cu^{2+}, and Zn^{2+} had no effect. The molecular weight of the enzyme, by gel filtration, was 75,000. The enzyme was very specific for phosphatidate. It had some activity towards monoacylphosphatidate, but no activity towards phosphatidylinositol 4-phosphate, phosphatidylinositol,

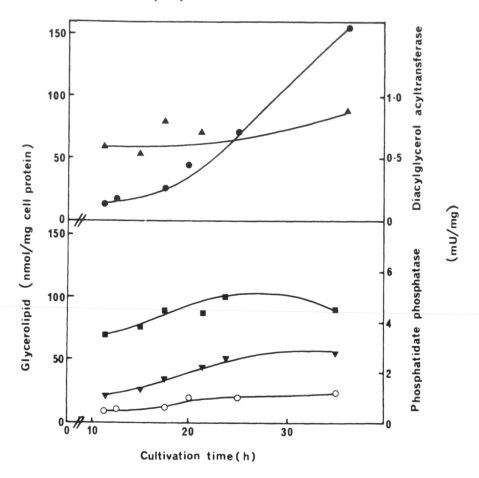

FIGURE 20.   Changes in diacylglycerol acyltransferase and phosphatidate phosphohydrolase activities
and lipid composition with cultivation time in *S. cerevisiae*. (O—O) Diacylglycerol, (●—●) triacylglycerol, (▲—▲) diacylglycerol acyltransferase, (■—■) membrane-bound phosphatidate phosphohydrolase, (▼—▼) soluble phosphatidate phosphohydrolase. Taken from Reference 97 where
further details can be found.

rac-glycerol 1-phosphate, glycerol 2-phosphate, 5′-AMP, 3′-AMP, phosphoserine, glucose
6-phosphate, or *p*-nitrophenylphosphate (Table 9).

Triacylglycerols accumulated in the stationary phase of growth of *S. cerevisiae* and incorporation of [^{14}C]acetate into triacylglycerols increased, while incorporation into phospholipids remained constant.[97] The accumulation of triacylglycerols could not be correlated
with a change in fatty acid synthesis. Measurement of soluble and membrane-bound phosphatidate phosphohydrolase activities showed that the activity of the soluble enzyme was
higher in the stationary than the exponential phase of growth. The activity of the membrane-
bound enzyme also increased, as did the glycerophosphate acyltransferase activity, but to a
lesser extent than a soluble phosphatidate phosphohydrolase which was about threefold higher
in stationary than early exponential phases of growth. The activities of cholinephosphotransferase and diacylglycerol acyltransferase (which require diacylglycerol as a substrate) remained relatively constant. Therefore, an increase in triacylglycerol accumulation was closely
correlated with an increase in phosphatidate phosphohydrolase activity, which gave a change
in the partitioning of phosphatidate between diacylglycerol (required for triacylglycerol
biosynthesis) and CDP-diacylglycerol (required for phospholipid biosynthesis)[97] (Figure 20).

The possibility was suggested that the soluble and membrane-bound forms of phosphatidate phosphohydrolase were two distinct proteins under different genetic control and that the soluble enzyme had a more important role in triacylglycerol biosynthesis.

It appears, therefore, that phosphatidate phosphohydrolase may have a regulatory role in lipid biosynthesis in yeast. Further purification of the enzyme will be necessary before this can be established.

C. Protozoa

Work with *Tetrahymena pyriformis* has shown that during the stationary phase of growth triacylglycerol biosynthesis is increased[98] in a similar manner to *S. cerevisiae*.[97] This also occurred under anaerobic conditions. However, phospholipid biosynthesis dropped in stationary phase and anaerobic growth conditions, whereas in *S. cerevisiae* phospholipid biosynthesis remained fairly constant.

In the stationary phase of growth the increase in triacylglycerol biosynthesis was accompanied by increases in the activities of acyltransferases and phosphatidate phosphohydrolase. In contrast, cholinephosphotransferase activity decreased.

Monoacylphosphatidate phosphohydrolase activity in the *Tetrahymena* cells was 3 to 11 times higher than phosphatidate phosphohydrolase activity. The monoacylphosphatidate phosphohydrolase activity was found mainly in the microsomal fraction and the activity was lower during the stationary phase and higher during the exponential phase of growth. Phosphatidate phosphohydrolase activity was mainly soluble and was higher during the stationary phase, when triacylglycerols were being accumulated. Substrate specificities of the two activities were not examined.

Shifting the cells to anaerobic conditions did not significantly alter the activity of either enzyme from that during the exponential phase of growth. However, as triacylglycerols are also accumulated under anaerobic conditions, but not during aerobic exponential growth, two separate mechanisms for control of triacylglycerol biosynthesis must exist.

Under anaerobic conditions, where no significant changes in activity of triacylglycerol biosynthetic enzymes occurs, there may be a shift in the utilization of diacylglycerol towards triacylglycerol biosynthesis and away from phospholipid biosynthesis, caused by a fluctuation in cellular levels of diacylglycerol. This suggests a possible regulatory role for phosphatidate phosphohydrolase, which is responsible for the production of the diacylglycerol.

A regulatory role is also possible for phosphatidate phosphohydrolase in the stationary phase of growth where an accumulation of triacylglycerols is accompanied by an increase in activity of phosphatidate phosphohydrolase in a similar way to *S. cerevisiae*.[97] However, there is also a possibility of regulation of triacylglycerol biosynthesis by monoacylphosphatidate phosphohydrolase, which has a higher specific activity than phosphatidate phosphohydrolase, but whose activity remains the same during stationary and exponential growth phases, particularly as *Tetrahymena* has an active 1-acylglycerol acyltransferase enzyme.

The role of phosphatidate phosphohydrolase and monoacylphosphatidate phosphohydrolase in *Tetrahymena* requires further study to see if there are two distinct enzymes with different substrate specificities. It is possible that there is one enzyme as in *Escherichia coli*,[90] and that it has different affinities for monoacylphosphatidate and phosphatidate which depend on whether it is present in the membrane-bound or soluble form. In addition, the phosphatase activity described may be due to a nonspecific phosphatase present in *Tetrahymena*, which is active at pH 7.2.

REFERENCES

1. **Mudd, J. B.,** Phospholipid biosynthesis, in *The Biochemistry of Plants,* Vol. 4, Stumpf, P. K. and Conn, E. E., Eds., Academic Press, New York, 1980, 249.
2. **Moore, T. S.,** Phospholipid biosynthesis, *Annu. Rev. Plant Physiol.,* 33, 235, 1982.
3. **Douce, R. and Joyard, J.,** Plant galactolipids, in *The Biochemistry of Plants,* Vol. 4, Stumpf, P. K. and Conn, E. E., Eds., Academic Press, New York, 1980, 321.
4. **Harwood, J. L. and Russell, N. J.,** *Lipids in Plants and Microbes,* Allen and Unwin, London, 1984.
5. **Harwood, J. L.,** Plant acyl lipids: structure, distribution and analysis, in *The Biochemistry of Plants,* Vol. 4, Stumpf, P. K. and Conn, E. E., Eds., Academic Press, New York, 1980, 1.
6. **Hitchcock, C. and Nichols, B. W.,** *Plant Lipid Biochemistry,* Academic Press, New York, 1971.
7. **Cline, K., Andrews, J., Mersey, B., Newcomb, E. H., and Keegstra, K.,** Separation and characterisation of inner and outer envelope membranes of pea chloroplasts, *Proc. Natl. Acad. Sci. U.S.A.,* 78, 3595, 1981.
8. **Block, M. A., Dorne, A.-J., Joyard, J., and Douce, R.,** Preparation and characterisation of membrane fractions enriched in outer and inner envelope membranes from spinach chloroplasts, *J. Biol. Chem.,* 258, 13281, 1983.
9. **Harwood, J. L.,** Plant mitochondrial lipids: structure, function and biosynthesis, in *Higher Plant Cell Respiration,* Douce, R. and Day, D. A., Eds., Springer-Verlag, Berlin, 1985, 37.
10. **Stumpf, P. K.,** Biosynthesis of saturated and unsaturated fatty acids, in *The Biochemistry of Plants,* Vol. 4, Stumpf, P. K. and Conn, E. E., Eds., Academic Press, New York, 1980, 177.
11. **Slabas, A. R., Harding, J., Hellyer, A., Sidebottom, C., Gwynne, H., Kessell, R., and Tombs, M. P.,** Enzymology of plant fatty acid synthesis, in *Structure, Function and Metabolism of Plant Lipids,* Siegenthaler, P.-A. and Eichenberger, W., Eds., Elsevier, Amsterdam, 1984, 3.
12. **McKeon, T. A. and Stumpf, P. K.,** Purification and characterisation of the stearoyl-acyl carrier protein desaturase and the acyl-acyl carrier protein thioesterase from maturing seeds of safflower, *J. Biol. Chem.,* 257, 12141, 1982.
13. **Roughan, P. G. and Slack, C. R.,** Cellular organisation of glycerolipid metabolism, *Annu. Rev. Plant Physiol.,* 33, 97, 1982.
14. **Joyard, J. and Stumpf, P. K.,** Characterisation of an acyl CoA thioesterase associated with the envelope of spinach chloroplasts, *Plant Physiol.,* 65, 1039, 1980.
15. **Joyard, J. and Stumpf, P. K.,** Synthesis of long-chain acyl-CoA in chloroplast envelope membranes, *Plant Physiol.,* 67, 250, 1981.
16. **Wharfe, J. and Harwood, J. L.,** Fatty acid biosynthesis in the leaves of barley, wheat and pea, *Biochem. J.,* 174, 163, 1978.
17. **Heinz, E. and Harwood, J. L.,** Incorporation of carbon dioxide, acetate and sulphate into the glycerolipids of *Vicia faba* leaves, *Hoppe-Seyler's Z. Physiol. Chem.,* 358, 897, 1977.
18. **Roughan, P. G., Mudd, J. B., McManus, T. T., and Slack, C. R.,** Linoleate and α-linolenate synthesis by isolated spinach chloroplasts, *Biochem. J.,* 184, 571, 1979.
19. **Jones, A. V. M. and Harwood, J. L.,** Desaturation of linoleic acid from exogenous lipids by isolated chloroplasts, *Biochem. J.,* 190, 851, 1980.
20. **Ohnishi, J. and Yamada, M.,** Glycerolipid synthesis in *Avena* leaves during greening of etiolated seedlings, *Plant Cell Physiol.,* 21, 1607, 1980.
21. **Sato, N. and Murata, N.,** Temperature shift-induced responses in the lipids in the blue-green algae *Anabaena variabilis, Biochim. Biophys. Acta,* 619, 353, 1980.
22. **Stumpf, P. K., Kuhn, D. N., Murphy, D. J., Pollard, M. R., McKeon, T., and MacCarthy, J.,** Oleic acid, the central substrate, in *Biogenesis and Function of Plant Lipids,* Mazliak, P., Beneveniste, P., Costes, C., and Douce, R., Eds., Elsevier, Amsterdam, 1980, 3.
23. **Marshall, M. O. and Kates, M.,** Biosynthesis of nitrogenous phospholipids in spinach leaves, *Can. J. Biochem.,* 52, 469, 1974.
24. **Sparace, S. A. and Mudd, J. B.,** Phosphatidylglycerol synthesis in spinach chloroplasts; characterisation of the newly synthesised molecule, *Plant Physiol.,* 70, 1260, 1982.
25. **Andrews, J. and Mudd, J. B.,** Characterisation of CDP-diacylglycerol and phosphatidylglycerol synthesis in pea chloroplast envelope membranes, in *Structure, Function and Metabolism of Plant Lipids,* Siegenthaler, P.-A. and Eichenberger, W., Eds., Elsevier, Amsterdam, 1984, 131.
26. **Mazliak, P. and Kader, J. C.,** Phospholipid-exchange systems, in *The Biochemistry of Plants,* Vol. 4, Stumpf, P. K. and Conn, E. E., Eds., Academic Press, New York, 1980, 283.
27. **Heinz, E.,** Enzymatic reactions in galactolipid biosynthesis, in *Lipids and Lipid Polymers in Higher Plants,* Tevini, M. and Lichtenthaler, H. K., Eds., Springer-Verlag, Berlin, 1977, 102.
28. **Van Besouw, A. and Wintermans, J. F. G. M.,** Galactolipid formation in chloroplast envelopes 1. Evidence for two mechanisms in galactosylation, *Biochim. Biophys. Acta,* 529, 44, 1978.
29. **Harwood, J. L.,** Sulpholipids, in *The Biochemistry of Plants,* Vol. 4, Stumpf, P. K. and Conn, E. E., Eds., Academic Press, New York, 1980, 301.

30. **Bishop, D. G., Sparace, S. A., and Mudd, J. B.**, Biosynthesis of sulfoquinovosyldiacylglycerol in higher plants: the origin of the diacylglycerol moiety, *Arch. Biochem. Biophys.*, 240, 851, 1985.

30a. **Kleppinger-Sparace, K. F., Mudd, J. B., and Bishop, D. G.**, Biosynthesis of sulfoquinovosyldiacylglycerol in higher plants: the incorporation of $^{35}SO_4$ by intact chloroplasts, *Arch. Biochem. Biophys.*, 240, 859, 1985.

31. **Harwood, J. L.**, The synthesis of acyl lipids in plant tissues, *Prog. Lipid Res.*, 18, 55, 1979.

32. **Singh, H. and Privett, O. S.**, Incorporation of ^{33}P in soybean phosphatides, *Biochim. Biophys. Acta*, 202, 200, 1970.

33. **Bieleski, R. L.**, Turnover of phospholipids in normal and phosphorus-deficient *Spirodela*, *Plant Physiol.*, 49, 740, 1972.

34. **Wilson, R. F. and Rinne, R. W.**, Studies on lipid synthesis and degradation in developing soybean cotyledons, *Plant Physiol.*, 57, 375, 1976.

35. **Smith, K. L., Douce, R., and Harwood, J. L.**, Phospholipid metabolism in the brown alga, *Fucus serratus, Phytochemistry*, 21, 569, 1982.

36. **Slack, C. R., Roughan, P. G., and Browse, J.**, Evidence for an oleoyl-phosphatidylcholine desaturase in microsomal preparations from cotyledons of safflower seed, *Biochem. J.*, 179, 649, 1979.

37. **Harwood, J. L.**, Radiolabelling studies of fatty acids in *Pisum sativum* and *Vicia faba* leaves at different temperatures, *Phytochemistry*, 18, 1811, 1979.

38. **Slack, C. R. and Roughan, P. G.**, The kinetics of incorporation in vivo of [^{14}C]acetate and $^{14}CO_2$ into the fatty acids of glycerolipids in developing leaves, *Biochem. J.*, 152, 217, 1975.

39. **Dorne, A.-J., Joyard, J., Block, M. A., and Douce, R.**, Localisation of phosphatidylcholine in spinach chloroplast envelope membranes, in *Structure, Function and Metabolism of Plant Lipids*, Siegenthaler, P.-A. and Eichenberger, W., Eds., Elsevier, Amsterdam, 1984, 339.

40. **Simpson, E. E. and Williams, J. P.**, Galactolipid synthesis in *Vicia faba* leaves. Sites of fatty acid incorporation into the major glycerolipids, *Plant Physiol.*, 63, 674, 1979.

41. **Roughan, P. G., Holland, R., and Slack, C. R.**, On the control of long-chain fatty acid synthesis in isolated intact spinach chloroplasts, *Biochem. J.*, 184, 193, 1979.

42. **Bertrams, M. and Heinz, E.**, Experiments on enzymatic acylation of *sn*-glycerol 3-phosphate with enzyme preparations from pea and spinach leaves, *Planta*, 132, 161, 1976.

43. **Joyard, J. and Douce, R.**, Site of synthesis of phosphatidic acid and diacylglycerol in spinach chloroplasts, *Biochim. Biophys. Acta*, 486, 273, 1977.

44. **Bertrams, M. and Heinz, E.**, Positional specificity and fatty acid selectivity of purified *sn*-glycerol 3-phosphate acyltransferase from chloroplasts, *Plant Physiol.*, 68, 653, 1981.

45. **Frentzen, M., Heinz, E., McKeon, T. A., and Stumpf, P. K.**, Specificities and selectivities of glycerol 3-phosphate acyltransferase and monoacylglycerol 3-phosphate acyltransferase from pea and spinach chloroplasts, *Eur. J. Biochem.*, 129, 629, 1983.

46. **Roughan, P. G., Holland, R., and Slack, C. R.**, The role of chloroplasts and microsomal fractions in polar lipid synthesis from [1-^{14}C]acetate by cell-free preparations from spinach leaves, *Biochem. J.*, 188, 17, 1980.

47. **Sumida, S. and Mudd, J. B.**, The structure and biosynthesis of phosphatidylinositol in cauliflower influorescence, *Plant Physiol.*, 45, 712, 1970.

48. **Kates, M.**, Hydrolysis of lecithin by plant plastid enzymes, *Can. J. Biochem.*, 33, 575, 1955.

49. **Blank, M. L. and Snyder, M. L.**, Specificities of alkaline and acid phosphatases in the dephosphorylation of phospholipids, *Biochemistry*, 9, 5034, 1970.

50. **Konigs, B. and Heinz, E.**, Investigation of some enzymatic activities contributing to the biosynthesis of galactolipid precursors in *Vicia faba, Planta*, 118, 159, 1974.

51. **Moore, T. S., Lord, J. M., Kagawa, T., and Beevers, H.**, Enzymes of phospholipid metabolism in the endoplasmic reticulum of castor bean endosperm, *Plant Physiol.*, 52, 50, 1973.

52. **Moore, T. S. and Sexton, J. C.**, Phosphatidate phosphatase of castor bean endosperm, *Plant Physiol.*, 61S, 69, 1978.

53. **Herman, E. M. and Chrispeels, M. J.**, Characteristics and subcellular localisation of phospholipase D and phosphatidic acid phosphatase in mung bean cotyledons, *Plant Physiol.*, 66, 1001, 1980.

54. **Van der Wilden, W., Herman, E. M., and Chrispeels, M. J.**, Protein bodies of mung bean cotyledons as autophagic organelles, *Proc. Natl. Acad. Sci. U.S.A.*, 77, 428, 1980.

55. **Sastry, P. S. and Kates, M.**, Biosynthesis of lipids in plants. II. Incorporation of glycerophosphate-^{32}P into phosphatides by cell-free preparations from spinach leaves, *Can. J. Biochem.*, 44, 459, 1966.

56. **Barron, E. J. and Stumpf, P. K.**, Fat metabolism in higher plants. XIX. The biosynthesis of triglycerides by avocado mesocarp enzymes, *Biochim. Biophys. Acta*, 60, 329, 1962.

57. **Joyard, J. and Douce, R.**, Characterisation of phosphatidate phosphohydrolase activity associated with chloroplast envelope membranes, *FEBS Lett.*, 102, 147, 1979.

58. **Krag, S. S., Robinson, M. D., and Lennarz, W. J.**, Biosynthesis of diacylglycerol from phosphatidic acid in the membranes of *Bacillus subtilis, Biochim. Biophys. Acta*, 337, 271, 1974.

59. **Block, M. A., Dorne, A.-J., Joyard, J., and Douce, R.,** The phosphatidic acid phosphatase of the chloroplast envelope is located on the inner envelope membrane, *FEBS Lett.,* 164, 111, 1983.

60. **Walker, K. A. and Harwood, J. L.,** Localisation of chloroplastic fatty acid synthesis *de novo* in the stroma, *Biochem. J.,* 226, 551, 1985.

61. **Shine, W. E., Mancha, M., and Stumpf, P. K.,** Fat metabolism in higher plants. The function of acyl thioesterases in the metabolism of acyl-coenzymes A and acyl-acyl carrier proteins, *Arch. Biochem. Biophys.,* 172, 110, 1976.

62. **Jamieson, G. R. and Reid, E. H.,** The occurrence of hexadeca-7,10,13-trienoic acid in the leaf lipids of angiosperms, *Phytochemistry,* 10, 1837, 1971.

63. **Heinz, E. and Roughan, P. G.,** Similarities and differences in the lipid metabolism of chloroplasts isolated from 18:3 and 16:3 plants, *Plant Physiol.,* 72, 273, 1983.

64. **Sato, N., Murata, N., Miura, Y., and Ueta, N.,** Effect of growth temperature on lipid and fatty acid compositions in the blue-green algae, *Anabaena variabilis* and *Anacystis nidulans, Biochim. Biophys. Acta,* 572, 19, 1979.

65. **Siebertz, H. P. and Heinz, E.,** Labelling experiments on the origin of hexa- and octadecatrienoic acids in galactolipids in leaves, *Z. Naturforsch.,* 32c, 193, 1977.

66. **Williams, J. P. and Khan, M. U.,** Lipid biosynthesis in *Brassica napus* leaves. 1. ^{14}C-labelling kinetics of the fatty acids of the major glycerolipids, *Biochim. Biophys. Acta,* 713, 177, 1982.

67. **Gardiner, S. E., Heinz, E., and Roughan, P. G.,** Glycerolipid labelling kinetics in isolated intact chloroplasts, *Biochem. J.,* 224, 637, 1984.

68. **Gardiner, S. E. and Roughan, P. G.,** Relationship between fatty acyl composition of diacylgalactosyl-glycerol and turnover of chloroplast phosphatidate, *Biochem. J.,* 210, 949, 1983.

69. **Williams, J. P., Khan, M. U., and Mitchell, K.,** Biosynthesis of monogalactosyldiacylglycerol in 16:3 and 18:3 plants: comparisons of diacylglycerol precursors and the effects of temperature on biosynthesis and desaturation, in *Structure, Function and Metabolism of Plant Lipids,* Siegenthaler, P.-A., and Eichenberger, W., Eds., Elsevier, Amsterdam, 1984, 123.

70. **Gardiner, S. E., Heinz, E., and Roughan, P. G.,** Rates and products of long-chain fatty acid synthesis from [1-^{14}C]acetate in chloroplasts isolated from leaves of 16:3 and 18:3 plants, *Plant Physiol.,* 74, 890, 1984.

71. **Clive, K. and Keegstra, K.,** Galactosyltransferases involved in galactolipid biosynthesis are located in the outer membrane of pea chloroplast envelopes, *Plant Physiol.,* 71, 366, 1983.

72. **Hilditch, T. P. and Williams, P. N.,** *The Chemical Constitution of the Natural Fats,* 4th ed., Chapman and Hall, London, 1964.

73. **Slack, C. R., Campbell, L. C., Browse, J. A., and Roughan, P. G.,** Some evidence for the reversibility of the cholinephosphotransferase-catalysed reaction in developing linseed cotyledons *in vivo, Biochim. Biophys. Acta,* 754, 10, 1983.

74. **Rochester, C. P. and Bishop, D. G.,** The role of lysophosphatidylcholine in lipid synthesis by developing sunflower seed microsomes, *Arch. Biochem. Biophys.,* 232, 242, 1984.

75. **Murphy, D. J., Woodrow, I. E., Latzko, E., and Mukherjee, K. D.,** Solubilisation of oleoyl-CoA thioesterase, oleoyl-CoA phosphatidylcholine acyltransferase and oleoyl phosphatidylcholine desaturase, *FEBS Lett.,* 162, 442, 1983.

76. **Stymne, S. and Glad, G.,** Acyl exchange between oleoyl-CoA and phosphatidylcholine in microsomes of developing soya bean cotyledons and its role in fatty acid desaturation, *Lipids,* 16, 298, 1981.

77. **Mancha, M., Garcia, J. M., and Quintero, L. C.,** Lipid synthesis in developing soybean seeds, in *Structure, Function and Metabolism of Plant Lipids,* Siegenthaler, P.-A. and Eichenberger, W., Eds., Elsevier, Amsterdam, 1984, 145.

78. **Stobart, A. K. and Stymne, S.,** The regulation of the fatty acid composition of the triacylglycerols in microsomal preparations from avocado mesocarp and the developing cotyledons of safflower, *Planta,* 163, 119, 1985.

79. **Stymne, S. and Appelqvist, L.-A.,** The biosynthesis of linoleate and linolenate in homogenates from developing soya bean cotyledons, *Plant Sci. Lett.,* 17, 287, 1980.

80. **Gunstone, F. D., Harwood, J. L., and Padley, F.,** Eds., *The Lipid Handbook,* Chapman and Hall, London, 1986.

81. **Stymne, S., Stobart, A. K., and Glad, G.,** The role of the acyl-CoA pool in the synthesis of polyunsaturated 18-carbon fatty acids and triacylglycerol production in the microsomes of developing safflower seeds, *Biochim. Biophys. Acta,* 752, 198, 1983.

82. **Gurr, M. I.,** The biosynthesis of triacylglycerols, in *The Biochemistry of Plants,* Vol. 4, Stumpf, P. K., and Conn, E. E., Eds., Academic Press, New York, 1980, 205.

83. **Vick, B. and Beevers, H.,** Phosphatidic acid synthesis in castor bean endosperm, *Plant Physiol.,* 59, 459, 1977.

84. **Frentzen, M., Hares, W., and Schiburr, A.,** Properties of the microsomal glycerol 3-phosphate and monoacylglycerol 3-phosphate acyltransferase from leaves, in *Structure, Function and Metabolism of Plant Lipids,* Siegenthaler, P.-A. and Eichenberger, W., Eds., Elsevier, Amsterdam, 1984, 105.

85. **Stymne, S. and Stobart, A. K.,** The biosynthesis of triacylglycerols in microsomal preparations of developing cotyledons of sunflower, *Biochem. J.,* 220, 481, 1984.

86. **Stymne, S. and Stobart, A. K.,** Evidence for the reversibility of the acyl-CoA: lysophosphatidylcholine acyltransferase in the microsomes of developing safflower cotyledons and rat liver, *Biochem. J.,* 223, 305, 1984.

87. **Griffiths, G., Stobart, A. K., and Stymne, S.,** The acylation of *sn*-glycerol 3-phosphate and the metabolism of phosphatidate in microsomal preparations from the developing cotyledons of safflower (*Carthamus tinctorius* L.) seed, *Biochem. J.,* 230, 879, 1985.

88. **Stobart, A. K. and Stymne, S.,** The interconversion of diacylglycerol and phosphatidylcholine during triacylglycerol production in microsomal preparations of developing cotyledons of safflower, *Biochem. J.,* 232, 217, 1985.

89. **Chang, Y.-Y. and Kennedy, E. P.,** Pathways for the synthesis of glycerophosphatides in *Escherichia coli, J. Biol. Chem.,* 242, 516, 1967.

90. **van den Bosch, H. and Vagelos, P. R.,** Fatty acyl-CoA and fatty acyl-acyl carrier protein as acyl donors in the synthesis of lysophosphatidate and phosphatidate in *Escherichia coli, Biochim. Biophys. Acta,* 218, 233, 1970.

91. **Weete, J. D.,** *Lipid Biochemistry of Fungi and Other Organisms,* 2nd ed., Plenum Press, New York, 1980.

92. **Kasinathan, C., Chopra, A., and Khuller, G. K.,** Phosphatidate phosphatase of dermatophytes, *Lipids,* 17, 859, 1982.

93. **Kuhn, N. J. and Lynen, F.,** Phosphatidic acid synthesis in yeast, *Biochem. J.,* 94, 240, 1965.

94. **Steiner, M. R. and Lester, R. L.,** *In vitro* studies of phospholipid biosynthesis in *Saccharomyces cerevisiae, Biochim. Biophys. Acta,* 260, 222, 1972.

95. **Cobon, G. S., Crowfoot, P. D., and Linnane, A. W.,** Phospholipid synthesis *in vitro* by yeast mitochondrial and microsomal fractions, *Biochem. J.,* 144, 265, 1974.

96. **Hosaka, K. and Yamashita, S.,** Partial purification and properties of phosphatidate phosphatase in *Saccharomyces cerevisiae, Biochim. Biophys. Acta,* 796, 102, 1984.

97. **Hosaka, K. and Yamashita, S.,** Regulatory role of phosphatidate phosphatase in triacylglycerol synthesis in *Saccharomyces cerevisiae, Biochim. Biophys. Acta,* 796, 110, 1984.

98. **Okuyama, H., Kameyama, Y., Yamada, K., and Nozawa, Y.,** Regulation of triacylglycerol and phospholipid synthesis in *Tetrahymena, J. Biol. Chem.,* 253, 3588, 1978.

Chapter 5

PULMONARY PHOSPHATIDATE PHOSPHOHYDROLASE AND ITS RELATION TO THE SURFACTANT SYSTEM OF THE LUNG*

Fred Possmayer

TABLE OF CONTENTS

* Abbreviations: DPPC, dipalmitoyl phosphatidylcholine; FPF, fibroblast pneumonocyte factor; L/S, lecithin/ sphingomyelin; NRDS, Neonatal Respiratory Distress Syndrome; PA, phosphatidic acid; PC, phosphatidyl- choline; PE, phosphatidylethanolamine; PG, phosphatidylglycerol; Pi, inorganic phosphate; PI, phosphatidy- linositol; PS, phosphatidylserine: RDS, Respiratory Distress Syndrome.

I. INTRODUCTION

It has been known from the time of Davson and Danielli and even before[1] that biological membranes contained amphipathic lipids. Phospholipids represent the principal constituents of cellular membranes and this accounts for their major function.[2] This being the case, one would anticipate that for the most part, the phospholipid composition of the major mammalian organs would exhibit similar patterns, both in terms of the relative amounts of the various phosphodiesters, and in terms of their individual molecular species. As has been well documented in a number of books and reviews,[3-6] this is far from the case. Furthermore, as indicated in the present volume, more than subtle differences have been noted in the metabolism of lipids in different tissues. Consequently, although the lung possesses approximately the same proportion of plasma membranes, mitochondria, and nuclear membranes as the other tissues featured in this book, it demonstrates a distinct phospholipid pattern,[3,6] implying, at least potentially, a different functionality. Of particular interest is the low proportion of highly unsaturated phosphatidylcholines and the high proportion of the disaturated species of this lipid. Copious amounts of DPPC have been observed in the lungs of all mammalian species studied including the mouse, rabbit, dog, cow, and monkey, as well as man.[6-9] In addition, significant amounts of DPPC have been observed in diverse species such as the chicken, turtle, and frog. The presence of DPPC in lung was first reported in 1944 by Thannhauser and colleagues.[10] Since that time this lipid has probably been subjected to more physicochemical examinations that any other lipid and perhaps as much as any other individual biological compound.

In addition to the presence of large amounts of DPPC, pulmonary phospholipids also show an even more pronounced difference from the lipids of other tissues in that this organ contains significant amounts of PG.[3,6,9] This lipid, which along with PI must be classified as a phosphorus-containing glycolipid, accounts for a major proportion of the glycerolipids in prokaryotes. Normally present only in trace amounts in eukaryotes, this lipid functions as an intermediate in the pathway for cardiolipin biosynthesis. However, even heart with its predominance of mitochondria and cardiolipin does not possess sufficient PG for detection by standard analysis.

The major and indeed the best known function of the lung is to provide the structural basis for oxygenating the blood and for removing carbon dioxide. Efficient gaseous exchange requires an enormous area; in man, this amounts to approximately 70 m², the size of a tennis court. At the site of exchange, exogenous air must be brought into intimate contact with large volumes of blood. In fact the lung has been referred to as an emulsion of air and blood.[11] Perhaps not surprisingly, this mixture exhibits considerable instability. Pulmonary surfactant is the stabilizing agent which nature has provided to stabilize this crucial emulsion. It is this material which accounts for a large proportion of the DPPC and most if not all of the PG present in lung.

The physical basis of the requirement for pulmonary surfactant arises from the properties of curved surfaces. According to the law independently formulated on theoretical grounds by the English physicist Younge and the French mathematician Laplace in the 17th century,[12] the pressure difference across a bubble blown in water is equal to two times the surface tension divided by the radius

$$\Delta P = 2\ \sigma/r \qquad (1)$$

where P = the pressure difference across the sphere (dyne/cm²), σ = surface tension (dyne/cm), and r = radius (cm).

With bubbles the size of our alveoli, the forces generated by the Younge-Laplace equation become considerable. Since the pressure difference across a bubble or an alveolus is inversely related to the radius, the force across an alveolus should increase when we exhale. If the surface tension in our alveoli were equal to that of water, which is 70 dyne/cm at 37°C, there would be a tendency for small alveoli to collapse into larger alveoli and for general alveolar collapse. Evidence for this phenomenon, known as atelectasis, will be presented in a later section of this review. In addition, the sucking of fluid from interstitial spaces, a phenomenon known as transudation, would be promoted. Fortunately, in the presence of pulmonary surfactant the surface tension in our lungs varies between 27 to 30 dyne/cm at total lung capacity and less than 1 dyne/cm at low lung volumes. The cyclical changes in surface tension at the air-liquid interface of the alveoli reduce the work required to ventilate our lungs and serve to stabilize our terminal airways and prevent transudation. In addition to its traditional role of providing diacylglycerol for the production of neutral lipid for energy storage and nitrogen-containing phospholipids for cellular membranes, pulmonary phosphatidate phosphohydrolase must function in the formation of phospholipids, particularly DPPC, for alveolar surfactant. The principal purpose of this review will be to discuss the nature and the role of the various phosphatidate phosphohydrolases present in the lung. It is assumed that for the most part the readers of this chapter will be lipid biochemists who are more acquainted with lipid metabolism in liver, adipose tissue, and brain and those individuals who are in the process of obtaining a working appreciation of the surfactant system of the lung. Consequently, the author will provide a brief introduction to the nature and the function of pulmonary surfactant. A considerable impetus towards the study of lung biochemistry and physiology has arisen from the realization that surfactant deficiency is a primary factor in the development of the NRDS, the major cause of perinatal morbidity and mortality in developed countries.[13] Present strategies aimed at minimizing and preventing this malady will be discussed. Our current understanding of the nature of the various phosphatidate phosphohydrolases present in lung tissue and their relation to phospholipid synthesis in general and surfactant production in particular will be summarized. Finally, the various

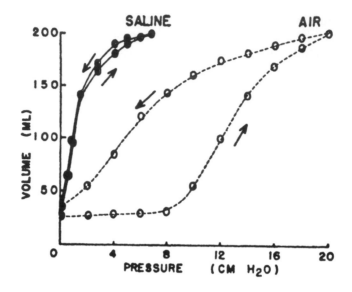

FIGURE 1.   Pressure-volume curves of excised cat lung inflated with air or saline. (These data are from Radford.[25] Van Neergaard only published the deflation sections for such curves.) (From Clements, J. A., *Proc. Int. Union Physiol. Sci.*, 1, 268, 1962. With permission.)

proposals for the control of glycerolipid synthesis in lung will be discussed in relation to the theories prevalent for liver and adipose tissue.

II. THE PULMONARY SURFACTANT: HISTORICAL CONSIDERATIONS

The first indication that the lung contains a surface-active material which facilitates lung expansion and stabilizes the terminal airways arose from pioneering studies conducted in the late 1920s by the Swiss physician von Neergaard.[14] von Neergaard examined the static retractive force of excised lungs, that is the natural tendency of the lung to contract, in a number of species including man. These studies revealed that the pressure required to maintain lungs inflated with air was considerably greater than that required to keep lungs inflated with isotonic solutions of gum arabic. He concluded that the difference between the pressures required to keep the lungs open with air compared to fluid must be due to surface tension forces (Figure 1). Since the static retractive force with air-filled lungs was two- to threefold greater than with fluid filled lungs, he concluded that the surface forces were more important than the forces needed to counteract tissue elasticity.

Using the methods of his day, which consisted of measuring the tendency of stretched films to contract, von Neergaard determined that gum arabic solutions recovered from the lungs produced surface tensions of 35 to 41 dyne/cm. These values were considered surprisingly low compared to the 60 dyne/cm observed with serum. von Neergaard postulated that "from a physicochemical point of view, it would be understandable that surface active substances gradually accumulate at the alveolar surface. It is also conceivable that this would be useful for the respiratory mechanism, because without it pulmonary retraction might become so great as to interfere with adequate expansion." Nevertheless he recognized the limitations of his methods and felt the matter required more study.

Although von Neergaard suggested that the phenomenon which he observed could prove important in the newborn, this work had no known impact on pediatric care or pathological investigation. In fact this provocatively simple study appears to have been completely ignored by pulmonary physicians and physiologists. As explained by Julius Comroe[15-17] in his fascinating historical account of "Premature Science and Immature Lungs", von Neergaard's discovery can be considered "premature" in that because of either a lack of comprehension or possibly a failure to appreciate its usefulness, this important concept was not incorporated into general scientific knowledge.[18] As a result, a fuller appreciation of the factors contributing to lung stability was delayed for 25 years. For those interested, an English translation of and a penetrating commentary on von Neergaard's manuscript are available.[19]

Histological evidence for the presence of a unique material in the alveoli was obtained by the studies of Charles C. Macklin who toiled at the same university as the author of this review. As early as 1946, Macklin recognized that the alveolar type II cells, which he referred to as epicytes or pneumonocytes, were secretory in nature and contributed some material to the fluid film which coated the alveolar walls.[20] He concluded that this material could be advantageous in external respiration. In 1954, after a forced retirement,[21] Macklin suggested that this mucoid-containing film could promote a number of vital functions including "the maintenance of a constant favourable surface tension, the facilitation of gaseous exchange, and the suppression of invading microorganisms".[22] Although these remarks were speculative, it is clear that Macklin recognized that the osmiophilic granules secreted from the granular pneumatocytes contained phospholipids which he referred to as "myelinogens".

A considerable part of Macklin's histological studies was supported by national defense funds obtained at the suggestion of Sir Frederick Banting of insulin fame. Banting was concerned with the obvious lack of scientific information applicable to chemical gas warfare. Unfortunately this meant that much of the histological data that Macklin produced during his prime remained sequestered in classified files even after the cessation of hostilities. Presumably, it was felt necessary to prevent the enemy from learning about the size of the alveolus. More direct evidence for the presence of a surface tension-reducing substance in the lung arose from two military scientists, R. E. Pattle at Porton in England and J. A. Clements at Edgewood in Maryland. Pattle, who had a longstanding interest in foams, was fortuitously asked to examine the pulmonary edema foam which was responsible for killing goats subjected to phosgene. He was surprised to observe that the bubbles produced under these conditions were remarkably stable to silicone antifoams. In 1955[23] he reported that bubbles obtained from cut rabbit lung or from tracheal washings were extremely stable. Bubbles produced in water or serum collapse quickly because the pressure difference generated by surface tension according to the Laplace equation, produces a continuous loss of gas across the surface. The remarkable stability of bubbles produced from the lung led Pattle to conclude they contained an insoluble material which reduced the surface tension to near zero. This material was susceptible to treatment with proteases. Pattle observed that the proposed substance was present in fetal rabbit lung at term. He suggested that the insoluble material produced a low surface tension in the alveolus which could be important in the prevention of transudation of fluid from interstitial spaces and blood capillaries.[24]

During the same period, other defense-supported pulmonary research conducted by Radford at Harvard was attempting to determine the alveolar surface area of mammalian lungs. Radford was attempting to establish the basis for a fuller understanding of gas diffusion into the terminal airways and of gaseous exchange into the capillaries. Because he suspected that histological estimates of alveolar size were exaggerated due to the fixation methods, Radford constructed pressure-volume loops (Figure 1) from which he could calculate surface free energy during deflation. Using an alveolar surface tension of 50 dyne/cm, the value observed with serum, he calculated that the surface areas obtained histologically were tenfold too high. Although he knew of von Neergaard's publication, Radford concluded that a sufficiently low surface tension to account for this difference was unlikely.[25]

John Clements, who was working at the Army Chemical Corps Medical Research Laboratories at Edgewood, Maryland, became uneasy about the tenfold variance discrepancy between Radford's data and the conventional values. When he applied the Younge-Laplace

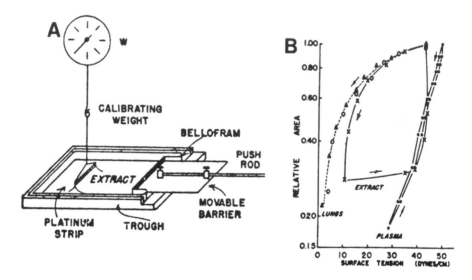

FIGURE 2.   (A) Diagrammatic representation of Clement's Langmuir-Wilhelmy surface balance. The surface tension of the monolayer is continuously monitored by the platinum strip suspended in the extract. (From Clements, J. A., *Proc. Int. Union Physiol. Sci.*, 1, 268, 1962. With permission.) (B) Surface tension-area curves obtained with saline extract of pulmonary surfactant. (From Clements, J. A., *Proc. Soc. Exp. Biol. Med.*, 95, 170, 1957. With permission.)

equation to this new information, Clements found that the calculated surface tension became much lower than 50 dyne/cm and furthermore that the surface tension varied with lung volume.

The realization that alveolar surface tension might change during inflation and deflation prompted Clements to design an apparatus which could examine surface tension during film expansion and compression. Using this dynamic system, which consisted of a modified Langmuir trough with its variable area modified to include a Wilhelmy hanging platinum plate to monitor surface tension (Figure 2A),[26] Clements was able to measure the surface tension of films produced from lung extracts. He discovered that the surface tensions of films from the lungs of rats, cats, or dogs fell from 45 dyne/cm to 10 dyne/cm or less during compression.[27] A marked hysteresis was observed during expansion when the maximum surface tension was rapidly achieved. The surface tensions observed with the surface tension balance during compression agreed well with those calculated for the lung (see Figure 2B).[28,29]

Further studies conducted by Pattle and Thomas,[30] Buckingham,[31] and Klaus et al.[32] concluded that phospholipids comprised the major surface tension reducing component of pulmonary surfactant and that the natural substance was a lipoprotein. Brown subsequently established that the disaturated lecithin, 1,2,dipalmitoyl-*sn*-phosphatidyl-3-choline (DPPC) was the major component.[33] As noted earlier, Thannhauser and his associates[10] had demonstrated the presence of DPPC in lung almost 20 years earlier, but this was the first indication of a special physiological function.

III. NRDS

NRDS has been known by at least a dozen different names. This is in keeping with Comroe's suggestion that the amount of information regarding the cause of a disease is inversely proportional to the square of the number of names acquired.[16] As early as 1903, Hochheim described layers of homogeneous materials which adhered to the walls of alveoli in infants succumbing shortly after birth.[34] Interest in these so-called "hyaline membranes" which were observed in many infants dying from neonatal insufficiency expanded rapidly from about the time of von Neergaard. Johnson and Myer,[35] who described these membranes in 1925, considered them to arise from aspirates of materials dispersed in amniotic fluid. The belief that the accidental aspiration of vernix (mainly sloughed off epithelial cells from fetal skin) essentially suffocated these newborn infants must have hampered a methodical examination of the genesis of this malady. Thus, the presence of an unexplained material in the lung clouded the issue so that the absence of pulmonary surfactant was not anticipated. In fact, because of the fear that feeding of prematurely delivered infants might result in the aspiration of gastric contents and thus create more membranes, such infants were compelled to survive on their meager nutritional stores.[36] It is true that at least one individual, Peter Gruenwald, felt that too much emphasis was being placed on hyaline membranes and suggested not only that the extensive atelectasis (alveolar collapse) associated with this disease was a primary factor, but also that a high surface tension was the underlying basis for its development.[37] Nevertheless, it was not until the 1950s that it was recognized that the so-called hyaline "membranes" were not truly membranes but a secondary effect due to tissue damage and transudation of serum proteins including fibrin. Hyaline membranes are thus bloodless clots that develop during the latter stages of this disease. These membranes are rarely found in infants succumbing during the first few hours of life.

The most common clinical picture[13] includes the initiation of respiratory difficulties in the first few hours after birth. The classical syndrome includes expiratory grunting, tachypnea (over 60 breaths per minute), and extremely "stiff" lungs as indicated by a resistance of the chest to expand to the extent that the sternum and intercostal ribs are clearly evident during inspiration. Cyanosis due to an inability to maintain adequate blood oxygenation, poor peripheral circulation, and a diffuse mottled appearance on X-rays, all attest to alveolar instability. The expiratory grunting appears to reflect the infant's attempt to prevent alveolar collapse by prolonging expiration and closing the glottis in order to prevent the pressure within the lung from falling to that of the external atmosphere. Current treatment includes not only elevated ambient oxygen but also mechanical means of maintaining a continuous positive pressure within the airways during mechanical or spontaneous ventilation.

In addition to hyaline membranes, NRDS has been attributed to asphyxia, heart failure, decreased lung volume, pulmonary hypoperfusion, disturbed autonomic regulation, deficiency of α-antitrypsin, and to the lack of fibrolytic enzymes. It is hardly surprising that this disease is still known as the Idiopathic (cause undetermined) Respiratory Distress Syndrome.

The rediscovery of the pulmonary surfactant system by Pattle and Clements soon inspired Avery and Mead to investigate the possibility that the absence of surfactant, as opposed to the presence of hyaline membranes, was responsible for the initiation of this disease in the neonate. Studies published by Avery and Mead in 1959[38] revealed that the lung extracts from infants dying of diseases other than respiratory distress displayed surface tension characteristics similar to those of extracts from older children or adults, while those succumbing to this disease produce extracts which exhibited minimal surface tension reduction or hysteresis. Minimum surface tensions in the high range (20 to 30 dyne/cm) were also observed with extracts from extremely premature infants weighing less than 1200 g. These findings, which were confirmed by a number of other studies, served to explain the high opening pressures required to inflate lungs from infants with NRDS, the lack of significant hysteresis between inflation and deflation curves, and the poor ability to maintain a residual volume in the absence of a distending pressure. They also emphasized the increasingly apparent relation between the incidence of this disease and premature delivery.

The recognition that infants dying of NRDS lacked sufficient surfactant to stabilize their alveoli has inspired a multidisciplinary approach to the study of this unique substance which has led both to an enormous acceleration in our scientific knowledge and to a rapid improvement in the methods applied to the management of the premature infant. Figure 3

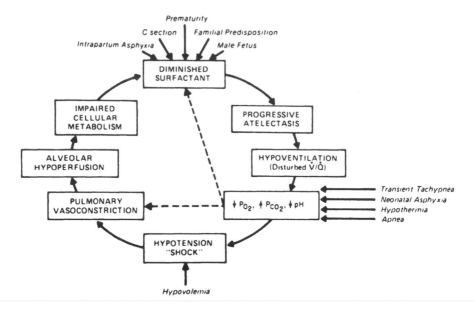

FIGURE 3.   Depiction of the "vicious cycle" in NRDS. Hypothetical and simplified representation of the factors presumed to play key roles in the pathophysiology of this disease. (From Farrell, P. M., in *Lung Development: Biological and Clinical Perspectives*, Vol. 11, Farrell, P. M., Ed., Academic Press, New York, 1982, 23. With permission.)

presents a summary of the principal abnormalities which appear to be involved in the "vicious" cycle which is produced by and/or which leads to NRDS.[13] As indicated, diminished surfactant can arise from a number of causes, the major determinant being prematurity. Male infants and infants of diabetic mothers are at higher risk. Although once considered an important determinant, Caesarean section does not appear to have a significant influence, except for its indirect effect related to prematurity. The factors involved in familial disposition probably involve pulmonary maturity relative to other organs at birth. There does not appear to be a genetic component to this syndrome. Diminished surfactant can lead inexorably to progressive atelectasis and thereby to ongoing reduction in pulmonary ventilation and adequate gaseous exchange. Inadequate gaseous exchange introduced through the above scenario or by neonatal asphyxia or transient apnea (pauses in breathing) can also lead to inappropriate systemic gases and blood pH. In the fetus, most of the blood coming from the heart bypasses the lungs via the ductus arteriosus and flows directly towards the brain. An alteration in blood gases *in utero* tends to promote constriction of pulmonary blood vessels. This produces a further decrease in pulmonary blood flow, thereby resulting in a sparing of blood for critical organs such as liver and brain. Premature infants show a disposition to retaining this fetal circulatory response to low blood oxygen. Pulmonary vasoconstriction can also be induced by hypotension due to low blood volumes. In either case, pulmonary vasoconstriction can result in a limited perfusion of those alveoli still being oxygenated. Not surprisingly, this physiological response, although providing survival value in the fetus, ultimately results in an impairment of cellular metabolic processes in the neonate, thereby further exacerbating the initial lack of sufficient surfactant. Evidence indicates that in the absence of sufficient surfactant the terminal bronchioles are more compliant than the terminal alveoli and so expand and contract during inspiration and expiration. The terminal bronchioles appear to be particularly fragile in the premature lung. The susceptibility of these terminal structures to the barotrauma produced during mechanical ventilation accounts for the leakage of serum materials leading to the formation of hyaline membranes in the latter stages of this disease. It should be clear that survival is dependent upon breaking this vicious cycle. As indicated by Farrell and Zachman,[13,39] current attempts towards preventing NRDS are directed towards the assessment of pulmonary maturation and the stimulation of pulmonary surfactant production in the fetus before delivery. The most rational management of this disease entails a combination of providing the missing surfactant and/or preservation of that surfactant available to the terminal airways.

IV. REDUCTION OF SURFACE TENSION BY PULMONARY SURFACTANT

The manner in which pulmonary surfactant stabilizes the lung by reducing the surface tension at the air-liquid interface of the alveoli will now be discussed in a semirigorous fashion. Surface tension arises from the cohesive properties of like molecules. A molecule of water in the bulk phase of a beaker experiences an attraction towards the water molecules around it with the result that the net forces cancel out. On the other hand, since molecules of water at the air-liquid interface experience negligible attraction towards air, the net result is an attraction downward into the bulk phase. Thus, all of the molecules at the surface experience a net attraction into the bulk phase. The most stable situation arises with a minimum surface area. One can think about surface tension in two ways. First, it takes work to move a molecule of water to the surface against the net downward attraction. The work required to expand the surface by one square centimeter at 37°C is 70 ergs. Likewise, one can think of surface tension in terms of the work one saves by maintaining surface area at a minimum. This is why droplets of liquids such as water and mercury, which have high surface tensions, form spheres. Thus, one can think of surface tension as a thin film at the surface which is resisting expansion at the surface or working towards minimizing it. It is this latter force or tension which is responsible for the pressure difference across a bubble described by the Younge-Laplace equation discussed earlier.

Polar lipids such as DPPC and unsaturated lecithins can be spread at the air-liquid interface so that the hydrophobic fatty acids extend into the atmosphere while the charged polar groups interact with the water (Figure 4). It will be recalled that surface tension arises from the disparity between the attractive forces on water molecules at the surface. Consequently, the binding of water molecules to the surface by charged polar groups results in a reduction in surface tension. It should be apparent that all phospholipids and in fact all surface active molecules can reduce surface tension in this manner. However, since phospholipids form insoluble films, compression of the surface monolayer, either by a barrier on a modified Wilhelmy-Langmuir surface balance or in the alveolus during expiration, results in an increase in the proportion of surface water bound and thus a further depression of surface tension. Phosphatidylcholines possessing short-chained fatty acids (e.g., dimyristoyl phosphatidylcholine) or phosphatidylcholines with double bonds have a limited ability to reduce surface tension. Such films are unstable and collapse during compression. On the other hand, films composed of DPPC, the major component of pulmonary surfactant, can be compressed until the surface tension falls to 0.7 dyne/cm or less at 37°C.[40,41] This reduction in surface tension by a factor of 100 compared to saline is sufficient to stabilize the alveoli and prevent alveolar collapse.

It should be noted that this simplified explanation of surface tension does not account for all of the observations related to surface tension. The reduction in surface tension by soluble substances such as ethanol is due to an increase in the relative concentration at the surface and thus a lessening of the absolute water concentration. (However, because ethanol is soluble, compression of the surface either with a barrier or through a decrease in the size of a bubble will have no effect on surface tension.[42]) It is likely that the failure to distinguish between the surface tension properties of soluble films produced by substances such as ethanol and insoluble films produced by substances such as proteins or phospholipids prevented von Neergaard from appreciating the truly remarkable surfactant properties of the

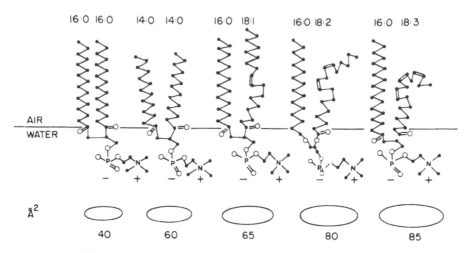

AVERAGE MOLECULAR CROSS SECTION OCCUPIED AT THE AIR-WATER
INTERFACE BY DIFFERENT PHOSPHATIDYLCHOLINES

FIGURE 4. Orientation of various molecular species of phosphatidylcholine at the air-liquid interface. The polar phosphorylcholine group is dissolved in the aqueous phase where it binds water by means of its positive and negative charges. Due to the "kinks" at the double bonds, phosphatidylcholines with unsaturated fatty acids cannot be compressed to the same extent as dipalmitoylphosphatidylcholine and therefore are not as effective in reducing the surface tension. Numerals refer to the number of carbons and the number of double bonds on the fatty acid: 16:0, palmitate; 14:0, myristate; 18:1, oleate; 18:2, linoleate; 18:3, linolenate. (From Possmayer, F., in *The Fetus and the Neonate*, Jones, C. T., Ed., Elsevier/North Holland, Amsterdam, 1982, 337. With permission.)

material he hypothesized was present in the alveolar lining layer.[16] In addition this simplified account does not resolve the basis of such related observations as the increase in surface tension due to high concentrations of sodium hydroxide.

V. COMPOSITION OF PULMONARY SURFACTANT

Pulmonary surfactant is routinely obtained by lavaging the lungs several times with saline, followed by centrifugation at low speeds to remove cells and at higher speeds to obtain a white pellet of crude surfactant.[43,44] This preparation can be further purified by relatively simple gradient techniques such as dispersion of the pellet with 0.75 *M* sucrose, overlaying with saline, and centrifugation at high speed in either a swinging bucket or a vertical rotor. Although it is possible to obtain various fractions of surfactant, a description of these procedures and the fractions produced[45] is beyond the scope of this review.

Despite some variety in the isolation methods, rather good agreement exists between the preparations recovered from a variety of species. Table 1 presents the composition of pulmonary surfactant from man,[47] from rat,[48] and from rabbit,[49] species which are most often used for experimental studies; and from bovine surfactant[50] which has been used in clinical trials. With all species DPPC constitutes the major component, varying between 35 to 40% of the total lipid. Unsaturated PC, which accounts for 25 to 35% in different preparations, is the next most abundant constituent. With surfactant obtained from adult lung, PG normally accounts for approximately 10%. As discussed further in another section, PG and PI have a reciprocal relation: PG is higher in the adult while PI is predominant in the fetal lung. Together these carbohydrate-containing acidic phospholipids make up to 10 to 12% of the total surfactant. Compositions for other species have been reviewed.[9,43,44] Surfactant obtained from birds and amphibians does not appear to possess PG, suggesting that this latter lipid was introduced late in evolution.[9,43] Although many authors have reported the presence of

Table 1
SURFACTANT COMPOSITION IN VARIOUS SPECIES (% TOTAL)

	Human		Rabbit			Cow
	Adult	Normal newborn	Adult	Newborn	Rat	Adult
Total lipid composition						
Neutral			9.0		10.4	3.0
Phospholipid			81.0		89.6	97.0
Phospholipid composition						
Phosphatidylcholine	80.5	77.0	91.6	79.2	81.0	79.2
(% saturated)	(44.7)					(59.2)
Lysophosphatidylcholine		0.7	trace	1.7	0.2	1.6
Phosphatidylethanolamine	12.3	4.9	1.2	4.7	4.6	3.5
Phosphatidylglycerol	9.1	6.0	2.7	4.2	5.0	11.3
Phosphatidylinositol	2.6	6.7	1.8	7.5	5.0	1.8
Phosphatidylserine	0.9	1.9		0.4		
Sphingomyelin	2.7	1.8	2.2	1.2	2.3	2.6
Lysobisphosphatidic acid		0.8		1.7		1.5
Others	1.9	0.2				
Phospholipid/protein ratio	8.3		11.2		8.8	9.0
Reference ( )	(47)	(94)	(48)	(49)	(48)	(50)

phosphatidylserine (PS) in surfactant from various species, studies in the author's laboratory have never observed this ninhydrin-staining lipid in purified samples of human, canine, rabbit, or bovine surfactant. All surfactant preparations contain significant amounts of PE. Small amounts of sphingomyelin (SM) and of an acidic lipid thought to be lyso-*bis*-phosphatidic acid are also observed.

Molecular species determinations on the phospholipids of pulmonary surfactant have concentrated on PC, the major component. In most species, the dipalmitoyl-species account for 60 to 70% of this lipid with monoenoic species making up the bulk of the remaining lecithin.[51-55] Studies on PG reveal a significant amount of disaturation, but not to the same extent as PC.[52-55] Perhaps because of the lesser amounts available, little work has been reported on the remaining phospholipids but total fatty acid compositions do not indicate marked disaturation. It is known that the PE fraction lacks significant quantities of disaturated species.[56]

Although virtually universal agreement exists regarding the phospholipids associated with pulmonary surfactant, considerable controversy accompanied the initial attempts at identifying the protein components intimately related to surfactant. The functional roles of the protein components of surfactant are still being investigated. Nevertheless, it has become clear that well-scrubbed preparations of pulmonary surfactant contain a major glycoprotein species of approximately 35 kdaltons and small amounts of protein with suggested nominal molecular weights of 6 to 10 kdaltons (Figure 5*).[57-59] The major component is usually accompanied by species with slightly lower and slightly higher molecular weights due to differences in carbohydrate content. In the absence of reducing reagents these proteins migrate on polyacrylamide gels as dimers and as more complex aggregates. Although it has been suggested that the 35-kdalton protein, sometimes referred to as "protein A", is derived from a macromolecular protein species known as "alveolin",[60] in vitro translation studies conducted with mRNA from human[61] and rat lung[62,63] demonstrate that the initial product is a 28-kdalton apoprotein which is subsequently modified by the addition of a complex asparagine-linked oligosaccharide to produce monomers with molecular weights of approximately

* Figure 5 follows page 54.

32 and 35 kdaltons. Two-dimensional isoelectric focusing gels reveal that each of the major molecular weight species consists of a train of individual peptides differing in the degree of sialylation and in the case of the rat in acetylation.[62-68] Surfactant glycoapoprotein A from man, cow, and rabbit possesses a single N-linked carbohydrate chain, while the surfactant apoprotein from the rat and dog appears to possess two glycosylated regions.

The cDNA for the dog glycoapoprotein A has now been cloned and sequenced by Benson et al.[69] These structures confirm earlier suggestions that a large section near the amino terminal corresponding to one third of the whole protein is composed of collagen-like sequences containing a series of glycine-X-Y amino acid triplets where Y is often hydroxyproline. The glycosylation site for dog glycoapoprotein A lies near the carboxy terminus. Using the canine cDNA as a probe, White et al.[70] screened a human genomic library to obtain the sequence for the human glycoapoprotein A gene. Comparison of the gene's sequence with that of the cDNA revealed an initiation site 24 to 37 nucleotides downstream from a consensus recognition sequence for initiation, TATAAT. The gene, which is nearly 5 kilobases long, is composed of five exons. The human cDNA sequence is characterized by a long (1000 nucleotides) 3'-untranslated region attached to the poly-A tail. A potential glucocorticoid binding site was located in close proximity to the glycoapoprotein A gene, suggesting that corticosteroids may be able to induce the production of this protein directly.

VI. PULMONARY DEVELOPMENT

A. Morphological Development

The mammalian lung originates from a branching endodermal tube arising from the foregut which invades a mass of mesenchymal cells derived from the splancho-pleural area of the embryo (see References 71 to 75 for further details). The development of the fetal lung can be divided into 3 stages: (1) a pseudoglandular stage, (2) a canalicular stage, and (3) a terminal sac stage. The pseudoglandular stage represents the growing of the primitive tracheobronchial tree into the loose mesenchymal tissue. As gestation advances, this endodermal outgrowth continues to branch asymmetrically, resulting in a structure with a glandular-like appearance. These "glands" are lined with undifferentiated columnar epithelial cells.

The canalicular phase is characterized by the opening of the "glands" to form ducts. This occurs at approximately 16 days in the rat (term 21 days) and 16 weeks (term 41 weeks) in the human.[71,75] The mesenchymal tissue now differentiates into interstitial tissue. Capillaries are formed in close approximation to the terminal respiratory bronchioles.

In the terminal sac stage, the epithelium of the terminal air spaces begins to differentiate into large thin type I cells and into the more rounded granular type II cells. This period was formerly known as the alveolar stage but its designation has been changed to acknowledge the fact that in many species, including the rat and man, alveolarization produced by the formation of alveolar septa subdividing the terminal air sac proceeds mainly after birth. The terminal sac period arises on day 21 in the rat and at approximately 25 weeks in the human. This final fetal stage leads into the final differentiation required to produce the competent structures required for airbreathing.

The type I cells cover most of the alveolar surface and possess a surface area approximately 50-fold greater than that of their type II counterparts. The former cells, which are highly differentiated and contain few organelles, appear well suited to act as the structural basis for gas diffusion to and from the nearby capillaries.

Although possessing less than half of the volume of their attenuated brothers, the type II epithelial cells have also been termed the "giant" alveolar cells. They possess a large nucleus, an extensive endoplasmic reticulum, comparatively large mitochondria, an extensive Golgi system, and multivesicular bodies. These latter structures are believed to function in the assembly of the osmiophilic lamellar bodies which are the hallmark of the differentiated type II cell (Figures 6A and B*).

The lamellar bodies, which represent stored surfactant, contain large amounts of phospholipid as well as protein and carbohydrate.[76-78] The appearance of large numbers of fully formed lamellar bodies signals the functional maturity of the fetal lung to the same extent as the presence of surfactant in the airways. Relatively pure lamellar bodies can be obtained from lung homogenates by gradient centrifugation. The chemical composition of this material is essentially identical to that of pulmonary surfactant obtained through lavage.[9,43,44] These characteristic organelles contain a large number of esterases and hydrolases possessing acidic pH optima.[79,80] This experimental evidence and the resemblance of their appearance under the electron microscope to the structures observed with lipidoses such as Tay-Sachs Disease indicate that the lamellar bodies represent modified lysosomes. These organelles contain alkaline as well as acid phosphatase and an apparently unique form of α-mannosidase.[81] Once secreted, the lamellar bodies can form a highly organized structure known as tubular myelin (Figure 7). It is this latter form of surfactant which appears to be responsible for the formation of the surface active monolayer.[45,82-84]

Evidence arising from toxicological studies reveals that the type I cells are considerably more susceptible to metabolic insults than are their type II counterparts. Type I cells do not appear to have the ability to divide; rather they are produced through a differentiation (or rather a dedifferentiation) process from type II cells. Type II cell proliferation occurs during recovery of the alveolus from a variety of injuries such as ozone damage, prolonged exposure to high oxygen, and treatment with butylated hydroxytoluene.[85,86]

B. Ontogeny of Pulmonary Surfactant

In addition to the morphological developmental stages described above, the final period of fetal pulmonary development can be divided into 4 functional periods which have been designated primitive, immature, transitional, and mature.[9] In most species the primitive period corresponds roughly to the pseudoglandular period and the immature to the canalicular period. The transitional period, which corresponds to the latter part of the canalicular stage and the initial part of the saccular stage, is identified as that gestational period when some but not all of the prematurely delivered fetuses will survive. The mature stage corresponds to that period where virtually all delivered fetuses will survive. In the rat, rabbit, sheep, and monkey the transitional period begins between 80 and 85% of gestation. The equivalent periods for the human are much earlier. In this case the transitional period can be considered as early as 26 weeks (65% of gestation) while, with the exception of infants born to diabetic mothers, few infants are lost after 32 weeks (78% of gestation). Only a small part of this difference can be attributed to artificial ventilation and the intensive care afforded by this species.

As indicated in Figure 8 the level of DPPC in human fetal lung increases rapidly between 50 and 75% of gestation (which corresponds more or less to the transitional period) but only between 85 to 90% in the other species considered.[87] Although we tend to consider ourselves a puny species, in terms of fetal lung maturity we are surprisingly robust. One can only speculate that this remarkably early maturation reflects an evolutionary adaptation which has arisen to reflect the extraordinary care which can be focused on our offspring.

VII. ASSESSMENT OF FETAL PULMONARY MATURITY

In addition to depicting the concentration of DPPC in fetal lung tissue, Figure 8 also reveals the level of DPPC in the alveoli of several species during gestation. It is apparent

---

* Figures 6A and 6B follow page 54.

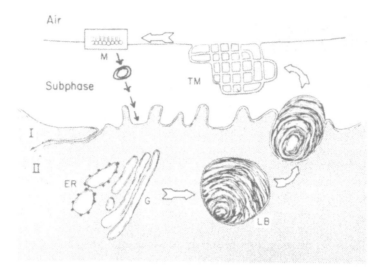

FIGURE 7.   Diagrammatic representation of the formation and excretion of pulmonary surfactant. After exocytosis, the lamellar bodies form tubular myelin which is thought to be the source of the monolayer. Only the monolayer can act in the reduction of the surface tension. ER, endoplasmic reticulum; G, golgi apparatus; LB, lamellar bodies; M, monolayer; TM, tubular myelin. (From Possmayer, F. et al., *Can. J. Biochem. Cell Biol.*, 62, 1121 1984. Modified from Goerke.[82] With permission.)

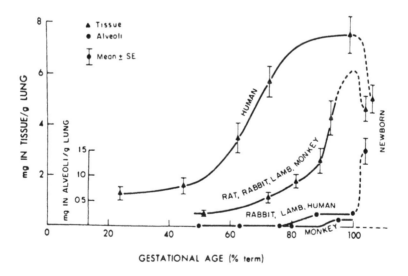

FIGURE 8.   Concentration of saturated PC in lung tissue and alveoli from different species plotted against relative gestational age. (From Clements, J. A. and Tooley, W. H., in *Development of the Lung*, Hodson, W. A., Ed., Marcel Dekker, New York, 1977, 349. With permission.)

that increasing amounts of DPPC are secreted near term and the amount of DPPC in the alveoli increases dramatically after birth. It has long been known from the classical tracheal ligation experiments of Jost and his Policard[88] that a considerable proportion of the amniotic fluid is derived from the fetal lung. The continuous flow of this fetal pulmonary fluid through the trachea continuously propels the alveolar contents towards a more susceptible pool. Since the lung often constitutes the most critical organ at birth, unambiguous indices of fetal pulmonary maturation are of particular value during this era of frequent elective preterm delivery. The reader is referred to recent reviews[89,90] of this area for specific details. Only a short synopsis can be given here.

A. Chemical Assays

Gluck and his associates[91] have established that the rapid increase in the level of PC in amniotic fluid is not accompanied by an increase in the level of sphingomyelin. Since sphingomyelin is predominantly derived from nonsurfactant sources, it can act as a useful internal standard to minimize the effects of variations in amniotic fluid volumes and to some extent the presence of nonpulmonary PC. The presence of sufficient PC to produce an L/S ratio of 2.0 virtually assures pulmonary maturity. Despite considerable progress in this area a significant number of false negatives (infants possessing a low L/S ratio who do not develop RDS) are still associated with this procedure. Fortunately, false positives (infants deemed mature who develop serious respiratory distress) are extremely rare. The false negatives may be attributable to variations in the storage/secretion ratio in the alveolus, the efficiency of alveolus to amniotic fluid transfer, and the extent of swallowing and the rate of degradation. Since the perinatal lung requires an adequate supply of surfactant regardless of the L/S ratio, the absolute level of PC in amniotic fluid also has predictive value and should be considered clinically along with the L/S ratio. The standard charring methods for densitometry do not char DPPC presumably because of its lack of unsaturated fatty acids. Techniques have been developed to overcome this deficiency.[92] Moreover, methods are also available for isolating disaturated PC from total lipids.[93] The basis of this separation involves complexing the double bonds of unsaturated fatty acids with osmium tetroxide, selectively eluting the un-reacted disaturated species through a short alumina column, and determining the quantities by phosphorus assay. While this test may remove ambiguity related to unsaturated vs. saturated PCs, the absolute PC level is readily available from the L/S ratio data and could be applied directly. This comment also applies to the palmitate/stearate and palmitate/oleate ratios in amniotic fluid since the absolute level of palmitate may also have predictive value.

The observation that infants recovering from RDS have PG in tracheal aspirates[94] led to the development of a more specific phospholipid test for pulmonary maturation. According to Kulovich and associates,[95,96] with normal pregnancies the level of PI as a percentage of total lipid peaks at 35 to 36 weeks gestation but then declines as PG appears (Figure 9). It should be stressed that while the percentage of phospholipid as PI declines, the absolute level of this lipid continues to increase. It has become clear that the determination of PG and PI provides a more accurate manner of predicting fetal lung maturity, especially with maternal diabetes.[89,90,96] However, many babies who are born before PG is detected in amniotic fluid still fail to develop significant signs of RDS.

The estimation of the surfactant-associated proteins by immunological means should provide, at least theoretically, a highly specific assay for determining pulmonary maturity. A number of publications have described mono- and polyclonal assays for the major 35-kdalton apoprotein.[97-99] It is anticipated that this test will gain wide acceptance in the near future.[100]

B. Physical Assays

The bubble stability or "shake test" developed by Schleuter et al.[101] involves shaking a small quantity of amniotic fluid in 50% ethanol and then observing whether a complete ring

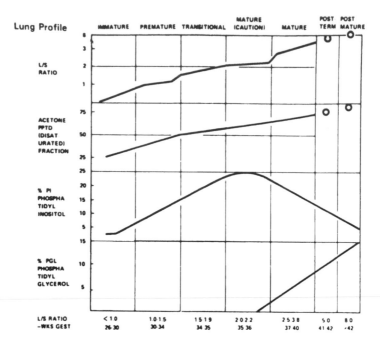

FIGURE 9. The normal lung profile showing the lecithin/sphingomyelin ratio, the proportion of PC which is precipitated by acetone, the percent PI and the percent PG in amniotic fluid samples at different weeks of gestation. (From Kulovich, M. V., et al. *Am. J. Obstet. Gynecol.*, 135, 57, 1979. With permission.)

of bubbles surrounds the inner edge of the tube. This assay is highly recommended by its ease and simplicity and by cost considerations, but the proportion of false negatives is still high. It serves as a useful screening test to reduce the number of amniotic fluids which must be examined chemically.

Fluorescence polarization (microviscosity) techniques have also been applied to amniotic fluids.[102-104] This method uses a fluorescent dye which dissolves in the lipid to detect the gel-fluid transition temperature of the bilayer. As the composition of the lipids in the amniotic fluid changes due to increased surfactant secretion, the decrease in the overall fluidity of the lipids can be precisely estimated. Although the instrumentation is expensive, the method is simple, rapid, accurate and requires very little sample. Nevertheless, while a few reports are available, general acceptance remains low. The microviscosity test could be influenced by neutral lipids and thus may be less specific for surfactant.

C. Enzymatic Assays

Many enzymes, particularly those of lysosomal origin, have been detected in amniotic fluid.[105] However, for the most part the activity profiles do not appear useful. The exception to this generalization is phosphatidic acid phosphohydrolase, the major topic of this review. The nature of the enzyme accumulating in amniotic fluid will be discussed in a latter section. Most studies have used [^{32}P]-labeled phosphatidate, but this substrate can only be prepared in relatively small quantities and has a short half-life. A highly sensitive assay based on a chemically synthesized [^{3}H]-labeled alkyl analog of lysophosphatidic acid should prove more useful in the clinical setting.[106]

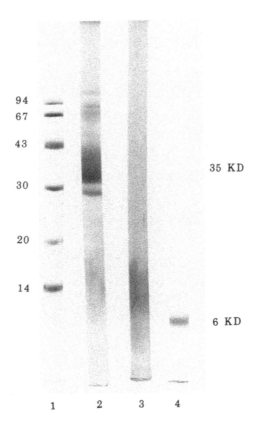

FIGURE 5. Polyacrylamide gel electrophoresis of bovine pulmonary surfactant under reducing conditions with sodium dodecyl sulfate. Lane 1, molecular weight standards. Lane 2. bovine pulmonary surfactant, showing the 35-kdalton apoprotein. The 6-kdalton apoprotein streaks and is difficult to visualize. Lane 3, Lipid extracts of bovine surfactant showing only 6-kdalton protein with lipid. Lane 4, Delipidated 6-kdalton apoprotein from pulmonary surfactant. (From Yu, S.-H., and Possmayer, F., *Biochem. J.*,232, 833, 1986. With permission.)

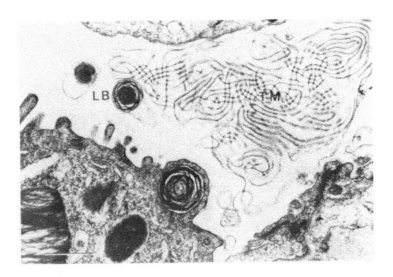

FIGURE 6A. Electron micrograph of alveolar type II cell showing lamellar bodies in cells and tubular myelin in the alveolar space. Some tubular myelin (TM) can be seen within the lamellar body (LB). (From Kuhn, C., III, in *Lung Development: Biological and Clinical Perspectives*, Vol. 1, Farrell, P. M., Ed., Academic Press, New York, 1982, 27. With permission.)

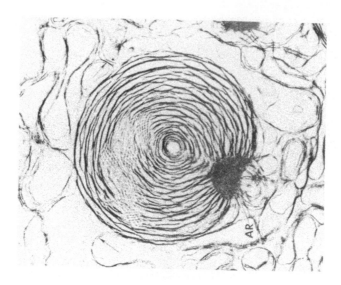

FIGURE 6B. Lamellar body from human amniotic fluid. The dark amorphous area (AR) is thought to contain mainly protein. Magnification × 33,600. (From Hook. G. E. R. et al., *Am. Rev. Resp. Dis.*, 117, 541, 1978. With permission.)

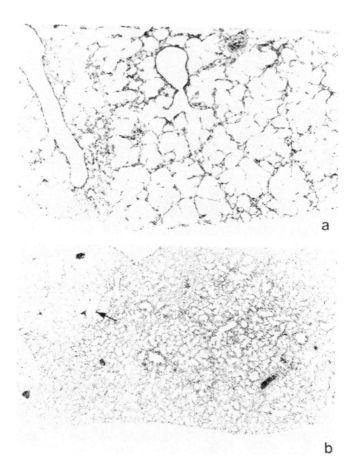

FIGURE 12. Effect of surfactant treatment on a prematurely delivered rabbit fetus (27 days gestation; term 31 days) on alveolar expansion. (a) Surfactant treated; (b) control littermate, magnification × 27. (From Robertson, B. and Enhorning, G., *Lab. Invest.*, 31, 54, 1974. With permission.)

VIII. ENDOCRINE AND PHARMACOLOGICAL REGULATION OF PULMONARY MATURATION

It has become clear that fetal pulmonary maturation can be accelerated by a number of hormones and other physiological agents.[107-112] In many cases the effects have been observed both in vivo and in vitro, indicating a direct effect on the lung. As yet, the direct biochemical effects of these agents have not been fully elucidated. At least two general mechanisms have been documented: an increase in the formation of pulmonary surfactant and a stimulatory effect on surfactant secretion. However, it is obvious that not all observations, for example morphological changes, can be attributed to the above alterations. The manner in which endocrine factors interact with the fetal and possibly the neonatal lung will likely continue to account for a major proportion of the scientific thrust in this area.

A. Glucocorticoids

During studies on the potential role of glucocorticoids in ovine parturition, Liggins[113] observed that these steroids precipitated not only the anticipated premature delivery but also produced precociously viable lungs. This serendipitous discovery has been followed by investigations with a large number of species including the rat, rabbit, sheep, and monkey which established that antenatal treatment at the appropriate gestation resulted in improved pressure-volume characteristics, enhanced surface activity of extracts, and elevated L/S ratios and morphological changes consistent with an advanced differentiation of the whole lung in general and the type II cell in particular.[107,109,111] These functional and morphological changes are accompanied by elevations in the level of PC in fetal lung and in the incorporation of radioactive choline and to a lesser extent other precursors into total and disaturated PC with slices of fetal lung.[107,111]

The effects of glucocorticoids are highly dependent on gestational age. Cortisol enhanced the growth of mixed cell cultures from fetal rabbit lung at 20 to 22 days gestation (term 31), but had little effect on cultures from 22 to 24 day fetuses and inhibited the growth of cultures from late gestation.[114] Cortisol accelerated glycogen deposition with alveolar type II cells from early gestation but promoted the depletion of glycogen from cells isolated near term. Incorporation of radioactive choline and palmitate into PC was stimulated only near term. Specific increases in the incorporation of radioactive precursors into disaturated PC have been reported for organ cultures from rat and human fetal lung.[115,116]

Fetal lung contains glucocorticoid receptors and studies in fetal rabbit lung indicate that these binding proteins function in the increase in PC production.[117] Although the possibility that these effects of glucocorticoids may reflect pharmacological rather than physiological responses has been considered, studies with the rat,[118] rabbit,[119] and human[120,121] are consistent with a temporal relationship between cortisol levels and maturation of the surfactant system. Interruption of the fetal pituitary axis by decapitation of rats or rabbits[122,123] *in utero* or inhibition of the endogenous steroid production with metyrapone[110] is also associated with a delay in lung maturation. However, under these circumstances and in the case of humans with congenital adrenal hyperplasia and even ancephaly, surfactant production eventually occurs, leading to the suggestion that endogenous cortisol may be a modulating rather than an essential factor. Considerable evidence has accumulated which indicates that endogenous cortisol production secondary to fetal stress can lead to enhanced pulmonary maturation.[124,125]

The realization that glucocorticoids can influence pulmonary maturation prompted Liggins and Howie[126] to design an extensive controlled double-blind clinical trial. This study demonstrated that if delivery can be delayed for 24 to 48 hr, maternal treatment with betamethasone leads to a marked reduction in the incidence and in the severity of RDS with infants of 28 to 32 weeks gestation. This encouraging initial study has been supported by a large number of other trials, in some cases less well controlled, which have nevertheless concluded that with the possible exception of pregnancies complicated by severe toxemia, adverse effects are minimal and considerable benefit appears likely.[127]

These encouraging results should be tempered by the obvious need for a bit of caution. The overall effect of glucocorticoids is to enhance cellular maturation. In many tissues including lung, liver, and brain the accelerated differentiation is accompanied by a marked decrease in cell division. Despite the large doses sometimes used in animal studies, it should be recognized that corticosteroids have multiple effects on dividing cells and that growing tissues are particularly susceptible. Given a complex tissue such as brain, a considerable abnormality could result from altering the normal pattern of multiplication, migration, and cell-cell interaction. A "catch-up" phenomenon has been observed with lung and brain.[128,129] Appropriate long term follow-up studies are still required, especially in the absence of a clear indication that the neonatal records of clinical units using steroids are significantly improved over those units using aggressive management of prematurely delivered infants.[130]

B. Mesenchymal-Epithelial Interactions

Glucocorticoids normally function through interaction with nuclear steroid receptors which influence the formation of specific messenger RNAs.[108] For the most part, experimental evidence demonstrating the direct formation of new proteins or the enhanced production of those proteins already present in type II cells is still not available. Interestingly, glucocorticoid treatment does not appear to induce the formation of the 35-kdalton surfactant apoprotein in either pure or mixed type II cell cultures.[448] Moreover, other evidence is available which indicates that the type II cells which produce surfactant may not necessarily be the most sensitive or the primary site of glucocorticoid action. Levels of dexamethasone which do not affect the number or the size of the lamellar bodies in fetal mice produce an increase in the size of the alveolar lumen.[131] Administration of betamethasone to pregnant rhesus monkeys resulted in an increase in maximal lung volume during the estimation of pressure-volume characteristics with prematurely delivered fetuses.[132] Since this effect was observed with saline-filled lungs where surface tension forces are minimal, the glucocorticoids appear to be acting on connective tissue, perhaps by affecting collagen-elastin relationship. It would be interesting to know whether these animals would survive. At present, the mechanism involved in the expansion of the future air spaces remains vague. One possible explanation is that the steroids accentuate the periods of fetal breathing.[133] It has become evident that fetal pulmonary fluid serves as an important molding factor in the development of the expanded alveoli.[134]

Mesenchymal interactions also play a role in mediating the actions of glucocorticoids on the surfactant system. Smith[109] has observed that while glucocorticoids have marked effects on type II cell activity in mixed cell cultures from fetal rabbits, only moderate increases in the incorporation of precursors into PC were noted with clones of human type II cells. This discrepancy has been explained through the presence of a fibroblast-pneumonocyte polypeptide factor (FPF) with a molecular weight of approximately 10 kdaltons. This factor, which is produced by lung fibroblasts in response to glucocorticoids, has the ability to stimulate surfactant production in vivo and in vitro.[109] Antibodies raised against this polypeptide prevent the increase in pulmonary maturation normally observed when pregnant rats are treated with glucocorticoids in late gestation.[135] The fetal fibroblast factor appears only to stimulate type II cell maturation and has no discernible effect on type II cells derived from adult animals. These observations are exciting not only because they provide a rational explanation for the coordinated development of mesenchymal and epithelial layers but also because they could provide the basis of a tissue-specific means of promoting surfactant synthesis without the potential deleterious effects of glucocorticoids.

C. Thyroid Hormone

Direct injection of thyroxine into rabbit fetuses produced an accelerated thinning of the alveolar septa, an advanced cytodifferentiation of the type II cells, and increased alveolar phospholipid levels and stability.[136] Treatment of pregnant rabbits with 3,5-dimethyl-3'-isopropyl-L-thyronine, a synthetic analog of triiodothyronine which crosses the placenta, resulted in an enhanced incorporation of choline into PC in fetal rabbit minces and elevated lung and alveolar levels of this lipid.[137] Thyroidectomy of fetal lambs in midgestation resulted in impaired alveolar differentiation, reduced surfactant secretion, and pronounced respiratory distress at term.[138] Recent studies indicate that the effects of glucocorticoids and thyroid hormone are superadditive.[115,139] Smith and Sabry[140] have suggested that the major effect of glucocorticoids is to stimulate mesenchymal production of FPF while thyroid hormone potentiates the effect of this factor on the type II cells.

The suggestion that thyroxine may play a role in pulmonary maturation and in the induction of surfactant synthesis is corroborated by the observation that the lung contains receptors for this hormone and that their level increases near term.[108,141] Infants with respiratory distress exhibit depressed serum thyroxine levels.[111] Clinical trials investigating the effect of this hormone in the human have produced variable results.[111]

D. Estrogen

A potential role for estrogen in the control of pulmonary maturation was first suggested by Abdul-Karim et al.[142] who proposed that in the rabbit 17β-estradiol may function in the development of the pulmonary vasculature. In the rabbit, maternal treatment with estradiol results in an enhanced incorporation of radioactive precursors, particularly choline, into PC.[107,111] Less evidence is available in other species. High levels of estrogen receptor have been observed in human and rat fetal lung.[143,144] Estrogens could act indirectly by promoting the secretion of prolactin. Although evidence has been presented indicating that this pituitary peptide does not function in the rabbit or sheep,[145-147] recent studies have indicated that it promotes the effects of glucocorticoids and thyroxine in explants of human lung.[148,149]

Recognition of estrogen as a factor in prompting fetal pulmonary maturity is of particular interest in reference to the well-recognized higher incidence of RDS in male infants.[9,109,111,112] Amniotic fluids from male infants exhibit lower L/S ratios and lower cortisol levels than females.[150] Male infants are also less responsive to glucocorticoid levels[127] and this may be mediated through the FPF system.[109] Fetal lung maturity is inhibited by androgens.[151]

E. cAMP

Considerable evidence has accumulated indicating that increasing the cAMP content of the lung either directly or indirectly, through the inhibition of phosphodiesterase by theophylline, produces a maturing effect on the lung in a number of species.[108-111] Treatment of pregnant rabbits at 27 days gestation with aminophylline (EDTA salt of theophylline) results in an increased incorporation of choline into disaturated PC, an elevation in fetal lung lavage levels, and an enhanced air retention during pressure-volume loops.[9] The presence of dibutyryl cAMP enhances the incorporation of palmitate and glycerol into PG with slices of fetal rabbit lung.[152] The decrease in glycogen levels in alveolar type II cells observed during the gestational stage when surfactant storage is accelerated[9,111] could be partially explained by cAMP effects on phosphorylase kinase. Glucocorticoids could function through the sensitization of adenyl cyclase or through the induction of β-adrenergic receptors.[111,153] In common with a number of peptide hormones, FPF could also function through an activation of adenyl cyclase.[109]

F. Insulin

It is now generally accepted that infants of diabetic mothers have a greatly increased risk of NRDS which is independent of gestational age or Caesarean section.[154] These infants display pancreatic islet hyperplasia and insulinemia, presumably related to maternal and consequently fetal hyperglycemia. Under these circumstances insulin acts as a growth-promoting factor in the fetus producing hypertrophy in muscle, adipose tissue, and possibly lung but delaying the functional maturation of the liver and lung. A number of studies with diabetic rabbits have revealed pressure-volume loops and surface balance results consistent with a delayed maturation of the surfactant system.[155,156]

The suggestion that insulin promotes pulmonary cellular growth while hindering differentiation is supported by the results of cell culture and slice experiments in the rabbit[157] and of explant studies in the rat[158] which suggest that in the presence of this hormone glycogen accumulates but the number of type II cells and lamellar bodies remains low. Insulin has a variable effect on the incorporation of radioactive precursors into total PC and a tendency to diminish the incorporation into the disaturated species.[157]

Several studies have suggested that insulin may hamper the corticosteroid-induced incorporation of precursors into PC and disaturated PC.[157,158] However it has also been reported that treatment of pregnant rabbits with cortisol reverses the functional delay of lung maturation due to alloxan diabetes.[159] In addition, studies with the rat indicate that maternal diabetes retards the normal increase in corticosteroids.[160] The manner in which these two hormones interact to affect lung maturation clearly requires more investigation.

Clinical interest has centered around the numerous observations indicating that with maternal diabetes an L/S ratio of 2.0 does not eliminate the possibility of NRDS.[89,90,96] In this situation, the rate at which the L/S ratio increases relative to term may not be greatly altered but the appearance of PG is retarded. Thus, further analysis of the amniotic fluid profile is required, especially with samples indicating marginal maturity.

IX. EFFECT OF HORMONES ON SURFACTANT SECRETION AND FLUID ADSORPTION

The continuous secretion of a portion of the surfactant stored in the alveoli during late gestation is markedly increased during labor.[111,161,162] This latter release is apparently related to elevations in fetal catecholamine levels.[163-166] It has become evident that the β-adrenergic agonists commonly used to suppress labor by inhibiting uterine contractions can also cross the placenta in sufficient quantity to promote secretion.[167,168] Studies with type II cells are consistent with the view that secretion can be promoted through β-adrenergic activation of adenylcyclase, thereby producing an increase in cAMP levels.[85,169] Microtubules and microfilaments appear to be involved in surfactant secretion. Actin acts as a major substrate for the cAMP-dependent protein kinase A in cytosols from adult rat lung or from alveolar type II cells.[170] The phosphorylation of actin, which is barely detectable in rat fetal lung in early gestation, increases somewhat near term and dramatically during the first week of life.[170,171] It has been shown that the level of β$_2$-adrenergic receptors in fetal lung increases near term and during the neonatal period in a number of species.[172-174] This increase can be accelerated by prior treatment with glucocorticoids, thyroid hormone, or estrogen.[175-179] Administration of an inhibitor of phenylethanolamine-*N*-methyltransferase to pregnant rabbits resulted in a decrease in epinephrine levels and a reduction in alveolar lipid after vaginal delivery.[180]

Air-breathing also augments surfactant secretion, but this appears to be controlled by another mechanism.[111,161,162] Evidence has been presented that cholinergic agents could also function to promote surfactant secretion through ventilation but this might reflect indirect effects.[111,162] In addition to β-agonists, it has been reported that surfactant secretion can be influenced by a number of diverse factors such as prostaglandins, phorbol esters, calcium ionophores, and metabolites of the expectorant, bromohexine. It is thought that some of these agents also promote surfactant synthesis.[168,181] This promises to be a fruitful area for future investigation.

It should be obvious that the transition to air-breathing requires a cessation of the secretion of pulmonary fetal fluid by the alveoli. In addition to their effects on surfactant secretion, β-adrenergic receptors also appear to be involved in the inhibition of pulmonary fluid outflow and in the initiation of its adsorption.[182-184] This reversal of alveolar fluid flow is related to an inhibition of chloride transport across the respiratory epithelium.[185] Alveolar type II cells isolated from fetal rabbits take up K^+ at 5 to 10% of the rate of adult cells. Labor stimulates both active and passive K^+ uptake so that fluid transfer approximates that in the adult lung.[186]

X. ASSAYS FOR PULMONARY SURFACTANT

As indicated in Section II, pulmonary surfactant was rediscovered by Clements through the use of a modified Langmuir-trough-Wilhelmy balance (Figure 2). This apparatus requires that the sample either be spread at the air-liquid interface or be adsorbed to the surface over a period of time. The resulting data give information concerning the compression or the respreading ability of the preparations. More recently, techniques have been introduced which examine the adsorption and spreading of surfactant at the air-liquid interface.[187-189] King and Clements[190] have introduced an adsorption assay which utilized a Wilhelmy platinum plate suspended in a large volume of saline medium in a Teflon beaker. A small sample of surfactant is injected into the subphase and the assay is initiated by activating a magnetic stir bar. The Wilhelmy plate monitors the surface tension with respect to time through a strain gauge. The end point achieved with this assay is approximately 27 dyne/cm. This corresponds to the equilibrium surface tension of a monolayer saturated with DPPC. In order to obtain lower surface tensions, it is necessary to compress the surface area either by a barrier as with the modified Langmuir trough or by decreasing the volume of a bubble. Samples of natural surfactant reduce the surface tension to approximately 27 dyne/cm within a few minutes. Pure lipids or the lipids extracted from surfactant produce only a slow moderate decline in surface tension with this apparatus.[190-193]

Studies in the author's laboratory have mainly been conducted with the pulsating bubble surfactometer introduced by Goran Enhorning.[194] This apparatus consists of a small sample chamber filled with approximately 25 μℓ of surfactant suspension. A bubble is created by drawing air through the small capillary which is in contact with the atmosphere (Figure 10). The bubble is mechanically pulsated between a radius of 0.55 and 0.4 mm at 20 cycles per minute. A pressure transducer monitors the pressure across the bubble. Since the radii are set and the pressure is monitored, the surface tension at any point can be calculated from the Laplace equation (Equation 1, Section I). The surface tension at maximum radius (R_{max}) and minimum radius (R_{min}) are normally expressed.

The pressure tracings obtained with saline, canine surfactant, and lipid extracts of canine surfactant are presented in Figure 11. With a bubble formed in saline, the surface tension remains at 70 dyne/cm regardless of the size of the bubble. When the bubble is created in a surfactant suspension, the pressure difference is much lower and the tracing is out of phase compared to the saline tracing. As the radius falls, the pressure decreases rather than increases. This indicates that the surface tension falls even more rapidly than the radius. When the bubble is at R_{max}, the surface tension is close to the equilibrium surface tension of DPPC, 27 dyne/cm. As the bubble approaches R_{min}, the surface tension falls to a value approximately 0 dyne/cm. As can be seen from Figure 11, similar surface tension characteristics are observed with lipid extract surfactant.

Comparison of the results obtained with these two assays revealed inconsistencies in the nature of the preparations and/or conditions required to obtain positive results (Table 2). A major difference was the amount of surfactant required. While 15 to 20 μg/mℓ of natural surfactant sufficed to produce a positive test with the Wilhelmy plate technique, 2.5 mg/mℓ were required with the pulsating bubble surfactometer. Lipid extract surfactant is active

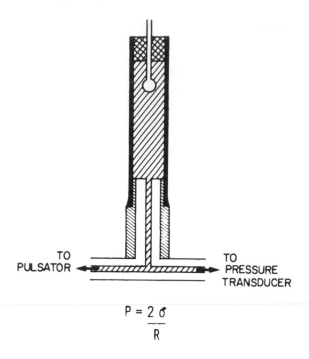

$$P = \frac{2 \, \sigma}{R}$$

FIGURE 10.   Diagrammatic representation of the pulsating bubble surfactometer. The bubble, which is in contact with the atmosphere, is continuously pulsated between radii of 0.4 and 0.55 mm. Surface tension is calculated from the recorded pressure by means of the Laplace equation. (From Possmayer, F. et al., *Can. J. Biochem. Cell Biol.*, 62, 1121, 1984. With permission.)

with the pulsating bubble but not with the Wilhelmy plate. EDTA at 3 mM selectively abolishes the activity observed with the Wilhelmy plate. A calcium requirement can be demonstrated with the pulsating bubble surfactometer, but this requires extracting this divalent cation by centrifugation through 50 mM EDTA three times. The activity is restored with calcium. Treatment of natural surfactant with trypsin or boiling for 5 min abolishes the activity observed with the Wilhelmy plate but has no effect on the pulsating bubble assay.

These observations suggested that the proteins associated with pulmonary surfactant might not be required for the activity monitored with the pulsating bubble.[195] However, treatment of natural surfactant or lipid extract surfactant with bacterial proteases or pronase reduced the effectiveness of these preparations on the pulsating bubble. In addition, although natural surfactant is stable to boiling for short periods, autoclaving at 121°C for 30 min or longer resulted in a reduction of the ability to reduce the surface tension to near 0 dyne/cm.

The differences in the requirements observed for the Wilhelmy plate and the pulsating bubble surfactometer led Possmayer et al.[189] to postulate that two separate mechanisms, termed the protein-facilitated process and the protein-mediated process, might be involved in the formation of the surface active monolayer. It was hypothesized that the protein-facilitated process, which can be readily observed with the Wilhelmy plate, requires low concentrations of lipid but native (undenatured) 35-kdalton apoprotein. This 35-kdalton protein-facilitated process also requires moderate levels of calcium. The 6-kdalton protein may also be required.

Treatment of natural surfactant with organic solvents, trypsin, heat, or EDTA produces an inactive preparation when tested with the Wilhelmy plate. These preparations, which

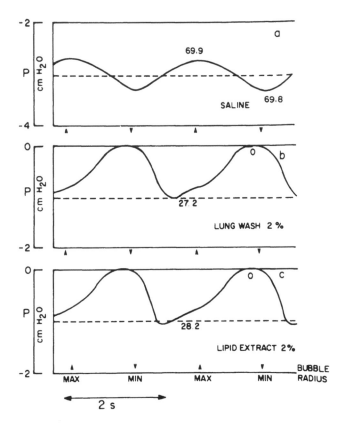

FIGURE 11. Pressure tracing (solid line) with (a) saline, (b) canine surfactant (2% wt/vol), and (c) lipid extract of canine surfactant (2% wt/vol) recorded with bubbles pulsating in the apparatus illustrated in Figure 10. The values along the tracing are surface tension in dynes × centimeter^{-1} at maximum and minimum bubble radii. (From Possmayer, F. et al., *Can. J. Biochem. Cell Biol.*, 62, 1121, 1984. With permission.)

remain active with the pulsating bubble surfactometer, do not possess the 35-kdalton apoprotein but retain the 6-kdalton apoprotein.[196-201] The requirement for the 6-kdalton protein in the protein-mediated process has recently been confirmed by reconstitution experiments.[198] Interestingly, once the 6-kdalton apoprotein is separated from the bulk of the PC it becomes susceptible to trypsin. Although earlier results indicated that the 6-kdalton apoprotein is a degradation product of its 35-kdalton counterpart, immunological studies indicate they are separate entities.[198,201]

Although lipid extract surfactant is not active with the Wilhelmy plate, these preparations were just as effective as natural surfactant in promoting expansion of lungs from prematurely delivered rabbit fetuses and in prolonging survival of such animals.[202-204] As will be discussed further in the next section, preparations of lipid extract surfactant have been used in clinical trials with prematurely delivered infants. These preparations contain only the 6-kdalton apoprotein. This indicates that the 6-kdalton protein-facilitated process is sufficient for at least some of the biological properties attributed to pulmonary surfactant. The reason that the lung possesses separate 35-kdalton and 6-kdalton mechanisms is not clear and the relation between these processes in the lung in vivo must still be determined.

In addition to its effects on formation of a surface active monolayer, the 6-kdalton apoprotein has been implicated in the control of surfactant uptake by alveolar type II cells[205] and in the regulation of surfactant lipid synthesis.[449]

Table 2
COMPARISON OF SURFACTANT ACTIVITIES ASSAYED WITH THE WILHELMY PLATE AND PULSATING BUBBLE TECHNIQUES

		Surfactant activity	
Preparation	Condition	Wilhelmy plate adsorption technique	Pulsating bubble surfactometer
Natural surfactant	15 μg/mℓ	+	−
Natural surfactant	2.5 mg/mℓ	(+)	+
Lipid extract		−	+
Natural surfactant	3 mM EDTA	−	+
Natural surfactant	3 mM EDTA, 5 mM CaCl$_2$	+	+
Lipid extract	10 mM EDTA	−	+
Natural surfactant	Trypsin treated	−	+
Natural surfactant	Boiled	−	+
Lipid extract	Boiled	−	+

Note: A positive test indicates the ability of a stirred preparation to reduce the surface tension to approximately 27 dyne/cm at 37°C within 1 min with the Wilhelmy plate or the surface tension of a pulsating bubble to 25 to 30 dyne/cm at R_{max} and 1 dyne/cm at R_{min} at 37°C. Unless indicated, the Wilhelmy adsorption technique employed approximately 15 μg surfactant/mℓ plus CaCl$_2$ and the pulsating bubble used 2.5 to 5.0 mg/mℓ surfactant based on phospholipid content. Results in parentheses are inferred from other available data.

From Possmayer, F. et al., *Can. J. Biochem. Cell Biol.*, 62, 1121, 1984. With permission.

XI. SURFACTANT REPLACEMENT THERAPY

Administration of exogenous surfactant appears to be the most direct approach towards preventing respiratory distress. This method would be particularly appropriate when rapid delivery prevents pharmacological acceleration of lung maturity. Surfactant administered as an iso-osmotic suspension before the first breath is layered onto the pulmonary fluid in the pharynx and drawn into the bronchioles during the first expansion. By lowering the surface tension at the air-liquid interface, surfactant greatly lowers the resistance to aeration. As pulmonary fluid is reabsorbed, the surfactant is retained on the alveolar walls where it continues to maintain alveolar stability during deflation. By minimizing the pressure gradient required to draw air into the alveoli, surfactant acts to prevent overextension of the terminal bronchioles during inhalation and their collapse during expiration.

The beneficial effects of treating prematurely delivered animals with pulmonary surfactant were first demonstrated by Robertson[168,181,206,207] using fetal rabbits of 27 or 28 days gestation (term 31 days). Deposition of homologous natural surfactant before the first breath resulted in a marked improvement in both the pressure-volume characteristics of the lungs[208-210] and in alveolar aeration as demonstrated by X-ray radiology[211] or microscopic examination either directly[212,213] or after fixation (Figure 12*). Surfactant treatment before the first breath resulted in prolonged survival of prematurely delivered rabbit[190,209] and rhesus monkey fetuses.[215] With both species, gaseous exchange during artificial ventilation was improved and bronchiolar epithelial necrosis markedly reduced or eliminated.[216,217] Similar overall findings have been reported from investigations on prematurely delivered lambs.[218-220]

The clinical trials that have been conducted can be divided into two major groups, those where the infants are treated prophylactically before the first breath and those where infants

* Figure 12 follows page 54.

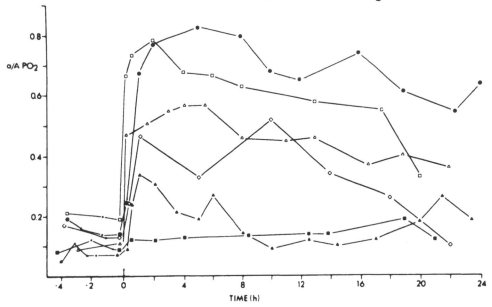

FIGURE 13.   Effect of surfactant treatment of prematurely delivered infants on the ratio of arterial to alveolar oxygen tension (a/A PO_2). Surfactant was given at time of 0. (From Smyth, J. A. et al., *Pediatrics*, 71, 913, 1983. With permission.)

with established NRDS are given surfactant to alleviate their difficulties in maintaining adequate gaseous exchange. Fujiwara and Adams[197,222] treated 28 air-breathing infants suffering from respiratory distress using a lipid extract from bovine surfactant to which DPPC and PG had been added. Since surfactant does not spread well in an air-filled lung, large amounts of surfactant (150 mg/kg) were administered and large volumes were used to recreate the first breath. This treatment resulted in improved gaseous exchange, normally permitting a marked reduction in ambient oxygen within a few hours. Improved pulmonary function was followed by clearing of chest radiograms, an increase in systolic blood pressure, appearance of diuresis, and the recovery of bowel movements. However, in many cases symptoms of patent ductus arteriosus ensued which required surgical correction.

Similar results have been observed in several other trials,[223-224] one of which used human surfactant prepared from amniotic fluid.[224] Figure 13 from studies conducted by Smyth et al.[223] shows the effect of surfactant on gaseous exchange in six infants suffering from advanced NRDS. The results are expressed as the ratio of arterial to alveolar PO_2. In healthy infants delivered at term this ratio is approximately 0.75. Surfactant treatment resulted in a marked improvement in four of the infants although the ratio tended to fall over the next 24 hr. In two cases very little or no improvement was noted. These two infants possessed the lowest arterial/alveolar PO_2 ratio prior to treatment. It is known that mechanical ventilation of the surfactant-deficient lung leads to damage of the terminal airways epithelium and transudation of serum proteins.[206,207,226] Serum contains a 110-kdalton protein which is particularly effective in inactivating pulmonary surfactant.[227] These results suggest that in some cases the surfactant was instilled into a pool of surfactant inhibitor protein and this counteracted the desired effect. These observations serve to emphasize the need for the rapid treatment of premature infants either prophylactically or in the early stages of NRDS. It should be noted that respiratory distress is a highly variable disease and these studies did not include controls. Nevertheless, the uniformity of the improvements in arterial oxygen reported by the different groups following surfactant insufflation indicates an obvious physiological response. However, there was no clear indication of an acceleration in the time course for resolving the premature lung disease for example by reducing the period required for mechanical ventilation.

The relative success of these initial "rescue operations" has been followed by other investigations in which prematurely delivered infants were treated prophylactically with surfactants derived by lipid extract of surfactant obtained from bovine surfactant, by human surfactant, and by artificial surfactants produced from synthetic and semisynthetic lipids. Bangham and associates[228,229] have adopted a highly original approach to preparing artificial surfactant for clinical use. This ingenious investigator observed that while liposomes of DPPC plus PG in a ratio of 7:3 were not particularly effective, dry flakes of this material rapidly reduced the surface tension of an air-liquid interface in a manner similar to dry flakes of natural surfactant. Apparently as the anhydrous lipids become hydrated, they adopt a nonliposomal open-ended configuration which can rapidly form a monolayer of DPPC and PG. The PG is squeezed out of the monolayer during compression. Treatment of prematurely delivered rabbit fetuses subjected to artificial ventilation revealed improved compliance compared to controls but only a minimal reduction in bronchiolar lesions and hyaline membranes.[230] In this study some lesions were also observed with natural surfactant, presumably because surfactant was not administered before the first breath.

In a very early prophylactic trial, Morley et al.[231] observed that infants of 34 weeks or less who were treated with aerosolized dry surfactant required considerably less ventilatory care than another group of untreated infants managed by the same unit. All of the treated group survived while 24% of the "control" group succumbed. The results indicated that treatment with dry surfactant could be beneficial but in the absence of a true control group the interpretation must be guarded. This caution is reinforced by reports from two separate groups using the same material which failed to observe a significant improvement in relatively small controlled studies.[232,233]

Durand et al.[234] have produced an artificial surfactant containing DPPC, hexadecanol, and a detergent tyloxapol. This approach to artificial surfactant involves maintaining the DPPC as a monolayer on the hexadecanol which acts as a carrier oil. The tyloxapol assists in the preparation of this mixture which is presently being used in controlled trials both prophylactically and in the treatment of infants with the established syndrome.

A considerable number of trials are currently in progress testing the effectiveness of bovine surfactant prepared following Fujiwara's[197,222] procedure which involves salt extraction of minced lungs followed by gradient centrifugation and lipid extraction. Human surfactant from amniotic fluid has also been utilized.[235] In addition, three trials have been reported in which lipid extract surfactant prepared from lung lavage of calves or young cows was given prophylactically to infants of less than 30 weeks gestation.[236-238] These studies uniformly revealed a marked decrease in the incidence and the severity of NRDS and a reduction in the complications normally associated with this disease. In the larger of these trials the requirement for excess oxygen defined as oxygen above room air[21%] was reduced by more than half and neonatal mortality fell from over 19% to less than 3% in the treated group. The impact of these results was somewhat mollified by the fact that the control group possessed a larger proportion of males than females. As indicated earlier, males are more susceptible to respiratory distress. Surprisingly, a greater number of fatalities was noted in the females than in the males of the control group. When taken together, the results of these three trials and the results of another study involving human surfactant provide strong evidence for the view that prophylactic treatment with surfactant has a beneficial effect on pulmonary function with these small babies and the potential for considerably reducing perinatal morbidity and mortality.

Because surfactant based on lipid extract of natural product contains protein, concern has been expressed about the possibility of adverse immunological reactions. Possibly because immature infants are not particularly immunocompetent, sera from treated infants treated with the Fujiwara preparation have not exhibited positive antibody reactions to date.[450] Nevertheless, preparations based on the natural product require considerable manual effort and are expensive. Each preparation must be tested individually for surfactant properties. This argument also applies to human surfactant. Since these preparations are derived from biological sources, potential problems related to sterility and, despite the results to date, to antigenicity, must still be considered. These biological and financial concerns highly recommend the use of artificial surfactants based on pure compounds. Although Bangham's dry surfactant and Clement's hexadecanol-based artificial mixture are not as surface active as lipid extract surfactant based on the natural material, if applied liberally and at the appropriate time they may prove sufficiently active to prevent respiratory distress in a manner similar to natural surfactant (see References 187 and 188 for a review of the critical surfactant properties). This remains undecided until a sufficient number of appropriately controlled investigations are conducted by a sufficient number of independent groups. In the interim, work is progressing on the cloning of the 6-kdalton apoprotein of pulmonary surfactant and the production of an artificial surfactant incorporating the human peptide.

XII. BIOSYNTHESIS OF GLYCEROPHOSPHOLIPIDS IN LUNG

It is anticipated that most of the readers of this review will have a working acquaintance with phospholipid biochemistry, so this section will be limited to a brief description of the overall pathways, stressing those studies giving a special insight into the mechanisms involved in the production of pulmonary lipids. Phospholipid synthesis in lung has been reviewed by a number of authors, and liberal use will be made of these previous publications.[51,107,239-245] Those readers requiring an introduction to the nomenclature of lipid chemistry are referred to the fine article by Sanders.[246] A more detailed account of the biosynthesis of many of the lipids discussed here can be found in the excellent text by Vance and Vance.[247]

B. Biosynthesis of Phosphatidylcholine

The initial step in the production of PC is the formation of PA, (reactions 1 to 5, Scheme 1). Studies, primarily in liver, have demonstrated that phosphatidate can be formed through the initial acylation of either *sn*-glycerol-3-phosphate or its more oxidized glycolytic precursor, dihydroxyacetonephosphate.[248,249] The primary acylation of dihydroxyacetonephosphate (reaction 2, Scheme 1) is followed by a reductive step (reaction 4) to yield the same intermediate, 1-acyl-*sn*-glycerol-3-phosphate, as would be formed through the direct acylation of glycerol-3-phosphate (reaction 3). The accumulation of 1-acyl-*sn*-glycerol-3-phosphate has not been documented in lung, presumably because this lysophosphatide is rapidly acylated to PA (reaction 5). The formation of the 1- rather than the 2-acyl isomer can be inferred from studies in liver.[248-250] The presence of 1-acyl-glycerol-3-phosphate acyltransferase has been well demonstrated.[248]

Initial studies in lung as in other tissues concerned the acylation of *sn*-glycerol-3-phosphate. Hendry and Possmayer[251] observed that while the microsomal fraction of rabbit lung possessed the highest specific activity of acyltransferase, the mitochondrial fraction also contained significant activity. In contrast to liver, where a strong preference for saturated fatty acids at the 1-position and unsaturated fatty acids at the 2-position of glycerol-3-phosphate is noted, considerably less specificity has been observed with microsomal preparations from lung. An overall lack of specificity has also been observed in the acylation of glycerol-3-phosphate with type II-like cells from pulmonary adenomas and with type II cells isolated from rat lung.[252,253]

The relative abilities of microsomes from rat lung and liver to specifically acylate glycerol-glucose →→

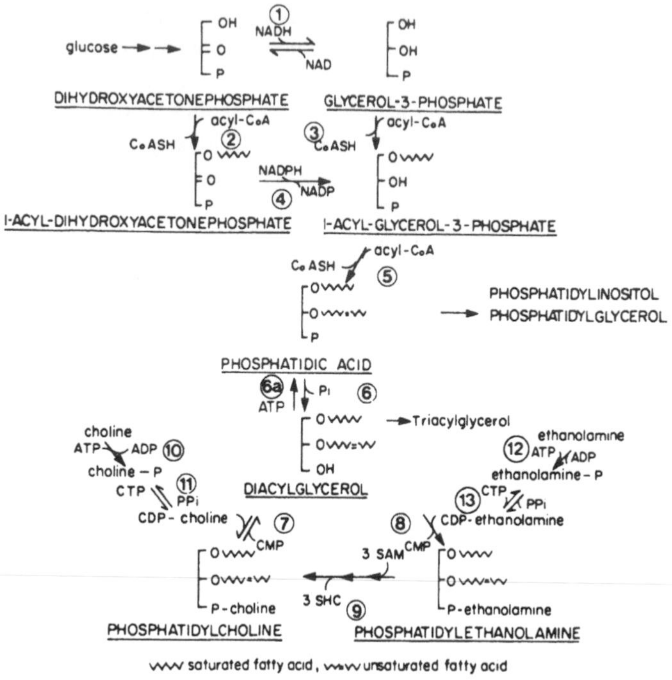

OH | O | P

① NADH / NAD

OH | OH | P

DIHYDROXYACETONEPHOSPHATE   GLYCEROL-3-PHOSPHATE

acyl-CoA ② CoASH

acyl-CoA ③ CoASH

O | O | P

NADPH / NADP ④

O | OH | P

1-ACYL-DIHYDROXYACETONEPHOSPHATE   1-ACYL-GLYCEROL-3-PHOSPHATE

CoASH acyl-CoA ⑤

O | O | P

PHOSPHATIDYLINOSITOL
→ PHOSPHATIDYLGLYCEROL

PHOSPHATIDIC ACID

⑥ₐ ATP / Pi ⑥

O | O | OH   → Triacylglycerol

choline
ATP / ADP ⑩
choline - P
CTP ⑪ PPi
CDP- choline CMP ⑦

DIACYLGLYCEROL

ethanolamine
⑫ ATP / ADP
ethanolamine - P
⑬ CTP / PPi
CDP- ethanolamine

⑧ 3 SAM CMP

O | O | P-choline

3 SHC ⑨

O | O | P-ethanolamine

PHOSPHATIDYLCHOLINE   PHOSPHATIDYLETHANOLAMINE

⌇⌇⌇ saturated fatty acid , ⌇⌇⌇ unsaturated fatty acid

DE NOVO SYNTHESIS OF PHOSPHATIDYLCHOLINE

Scheme 1.

3-phosphate were examined by Yamada and Okuyama.[254] Lung microsomes exhibited a greater capacity for incorporating palmitate at the 2-position. Type II cells possess a greater capacity for the acylation of glycerol-3-phosphate than whole lung.[253,255] As in the case of whole lung, there was little selectivity for palmitate over oleate at the 1-position or for oleate over palmitate at the 2-position. Although the results are consistent with the ability of these cells to produce disaturated PCs, the lack of specificity at the 1-position is difficult to rationalize with the data in vivo.

Studies conducted mainly in liver have revealed that in addition to microsomes, mitochondria and peroxisomes also possess significant primary acyltransferase activity.[249] The enzymes present in these three fractions can be distinguished on the basis of their substrate specificities and their susceptibilities to inactivation by heat and sulfhydryl reagents. While the microsomal enzyme can utilize either glycerol-3-phosphate or dihydroxyacetonephosphate, the mitochondrial enzyme only acylates glycerol-3-phosphate while the peroxisomal counterpart is specific for dihydroxyacetonephosphate. Although the subcellular distribution has not been extensively studied in lung, studies with radioactive precursors indicate that the endoplasmic reticulum acts as the major source of newly synthesized phospholipid for incorporation into lamellar bodies.[253,256,257] A possible involvement of mitochondria or peroxisomes has been suggested by studies with guinea pig fetal lung.[258]

Attempts to determine the relative importance of the glycerol-3-phosphate and the dihydroxyacetonephosphate pathways in the formation of PA have led to considerable controversy.[248,249] The most common approach has been to examine the relative incorporation of [2-^{3}H]glycerol and [^{14}C]glycerol into lipids with slices of the tissue being examined. The radioactive glycerols are taken up by the tissue and phosphorylated by glycerol kinase to give glycerol-3-phosphate with the same isotopic ratio as the extracellular precursor. However, since the tritium at the 2-position is lost when [2-^{3}H]glycerol-3-phosphate is oxidized to dihydroxyacetonephosphate, the ^{3}H/^{14}C ratio of this phosphorylated ketone falls to 0.

Theoretically, the isotopic ratio in the lipids should be the same as in the glycerol applied or 0, depending on whether lipids were synthesized via the glycerol-3-phosphate or dihydroxyacetonephosphate pathways respectively. A value falling between the original isotopic ratio and zero would indicate that both pathways were being utilized. In practice, the isotopic ratio observed was somewhat greater than in the administered glycerol. The basis of this unanticipated phenomenon was traced to the presence of an isotopic discrimination by glycerophosphate dehydrogenase against molecules containing tritium at the 2-position of glycerol-3-phosphate. The net effect is that molecules containing [^{14}C]glycerol-3-phosphate are preferentially converted to dihydroxyacetonephosphate. These molecules are taken up by the mitochondria and oxidized to CO_2 by the Krebs cycle. The net effect is that those molecules which contain tritium accumulate as glycerol-3-phosphate and the $^3H/^{14}C$ ratio increases. Using this approach Manning and Brindley[259] observed a sixfold increase in the isotopic ratio of intracellular glycerol-3-phosphate with slices of rat liver. With this preparation, the isotopic ratio of the lipids increased approximately twofold. Since the isotopic ratio was between that of intracellular glycerol-3-phosphate and zero, it is clear that both pathways are involved. Detailed analysis over the time period of the experiment revealed that the dihydroxyacetonephosphate pathway was responsible for 55% of the radioactive lipid being synthesized while the glycerophosphate pathway accounted for 45%.

The major criticism to the above approach is that the isotopic ratio of total cellular glycerol-3-phosphate may not be indicative of that pool which is being used to synthesize glycerolipids on the endoplasmic reticulum. A novel approach toward removing this ambiguity was adopted by Mason,[260] who examined the isotopic ratios in the headgroup glycerol of phosphatidylglycerol produced in type II cells in culture. Since the reaction which introduces the headgroup glycerol is specific for glycerol-3-phosphate, this value gives the isotopic ratio for the pool of glycerol-3-phosphate which is being utilized for phospholipid synthesis. Calculations based on applying this isotopic ratio to the isotopic ratios of the diacylglycerol glycerols in PG, PC, and DPPC concluded that the dihydroxyacetonephosphate pathway was utilized for 55% of the synthesis of these molecules.

PA produced by either of the above pathways does not accumulate in lung, where it accounts for less than 1% of the total lipid. The utilization of PA for acidic phospholipids will be discussed in another section. PA is hydrolyzed to diacylglycerol and Pi by phosphatidate phosphohydrolase before it can be used for the formation of PC and PE. The properties of phosphatidate phosphohydrolase in lung will be discussed in Sections XIII to XV.

It should be noted that in addition to the acylation mechanisms indicated above, PA can be formed through the phosphorylation of diacylglycerol with ATP by the enzyme diacylglycerol kinase (reaction 6A, Scheme 1). This enzyme which catalyzes the reversal of phosphatidate phosphohydrolase, has been studied in rat lung by Ide and Weinhold.[261] These authors observed that exogenous diacylglycerols containing unsaturated fatty acids were preferentially utilized. Diacylglycerol kinase acts as an important enzyme in the PI cycle for the production of inositol trisphosphate, an intracellular messenger involved in the control of intracellular Ca^{2+} levels.[262] The diacylglycerol produced via this cycle could also influence protein kinase C activity in some tissues. This latter enzyme has been implicated in surfactant secretion.[263] However, evidence for a direct role for the PI cycle in surfactant metabolism awaits further investigation.

The terminal reaction in the formation of PC is catalyzed by cholinephosphotransferase (reaction 7, Scheme 1). This enzyme has been extensively studied in lung preparation from a number of species (see References 9, 245, and 264). It is generally agreed that the pulmonary enzyme is located in the endoplasmic reticulum and requires Mg^{2+}. However, studies with fetal guinea pig lung are consistent with the view that other organelles such as microbodies could be involved.[265] Although cholinephosphotransferase and ethanolamine

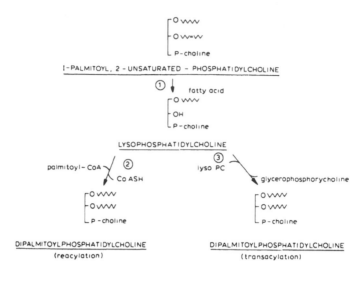

AUXILIARY PATHWAYS FOR THE FORMATION OF
DIPALMITOYLPHOSPHATIDYLCHOLINE

Scheme 2.

phosphotransferase share a number of properties, differences in the susceptibility to heat and proteolytic digestion indicate that separate peptides are required.[451]

Early studies with cholinephosphotransferase utilized exogenous 1,2-dipalmitoyl-*sn*-glycerol. Subsequent investigations revealed that under the experimental conditions, this disaturated substrate was not readily utilized by the microsomal enzyme compared to diacylglycerols containing unsaturated fatty acids (see References 9, 245, and 264). Recently, Miller and Weinhold[266] were able to demonstrate not only that disaturated diacylglycerols dispersed with PG can be utilized by this enzyme, but also that specific activities comparable with unsaturated diacylglycerols can be achieved in this manner.

Although significant amounts of PC are produced by the sequential methylation of PE with *S*-adenosyl methionine in liver[248] (reaction 9 a, b, c, Scheme 1), it has generally been concluded that this pathway is of marginal significance in lung. However, it should be noted that most of the experimental evidence involves studies examining the incorporation of ethanolamine through PE into PC. The relatively large pool size of PE would produce a large isotope dilution. Nevertheless, the lack of any increase in the incorporation of ethanolamine into PE in slices of fetal lung during the later stages of gestation indicates that this pathway does not play a major role in the form of PC for surfactant. It has recently been observed that the methylation pathway for PC synthesis is increased in lung during choline deficiency.[267]

B. Auxiliary Pathways for the Production of DPPC

It is generally accepted that in most tissues, including lung and liver, both glycerol-3-phosphate and dihydroxyacetonephosphate acyltransferase preferentially incorporate saturated fatty acids into the 1-position.[9,107,244,245] The second mole of palmitate could become associated with the 2-position of DPPC through the *de novo* pathway. However, initial investigations suggested that auxiliary mechanisms were involved in remodeling unsaturated phosphatidylcholines to DPPC. These auxiliary pathways include: (1) a deacylation-reacylation cycle in which the unsaturated fatty acid is removed from the 2-position of PC by phospholipase A_2 and the resulting 1-palmitoyl-lyso-PC is specifically reacylated with palmitoyl-CoA (reactions 1, 2; Scheme 2); (2) a deacylation-transacylation cycle, in which palmitate at the 2-position of one molecule of 1-acyl-lyso-PC is transferred to the 2-position of another lysophosphatide (reactions 1, 3; Scheme 2); and (3) a novel reaction in which either free or esterified palmitate is enzymatically exchanged with an unsaturated fatty acid at the 2-position of intact PC.

Both mechanisms a and b require the presence of a phospholipase A_2 type activity. Phospholipase A_2 activities have been examined in lung wash and in the cytosolic fraction of whole lung.[268-270] A Ca^{2+}-dependent phospholipase A_2 activity has been purified from the particulate fractions of rabbit lung.[271] This latter enzyme is more active with PE than with PC. The Ca^{2+}-dependent hydrolyses of PE by homogenates of fetal rat lung increased during gestation.[272] A Ca^{2+}-dependent phospholipase A_2 activity which exhibits specificity for unsaturated fatty acids at the 2-position of PC has been detected in rat lung microsomes.[273] Phospholipase A_2 activity has also been detected in pulmonary lysosomes, lamellar bodies, and in lung macrophages.[274,275]

The observation that the incorporation of 1-[^{14}C]palmitoyl-[^{3}H]lyso-PC into disaturated PC by rat lung slices was accompanied by a 2.5-fold increase in the ^{14}C/^{3}H ratio led to the suggestion that lyso-PC:lyso-PC transacylase (mechanism b) was involved in the production of surfactant DPPC.[51,276] This reaction was first described by Erbland and Marinetti.[277] The increase in isotopic ratio can be attributed to the fact that the radioactive lyso derivative, which was prepared from biosynthetically produced PC, contained 1-stearoyl-[^{3}H]lyso-PC molecules which are not reacylated as effectively as 1-palmitoyl containing lyso-PCs.[278]

It has now become apparent that the transacylation enzyme does not play a physiological role in surfactant synthesis.[9,245] The transacylase activity appears to be a minor enzymatic activity of lyso-PC hydrolase.[279] The transacylase as opposed to the hydrolase activity is observed at sufficiently high concentrations of 1-palmitoyl-lyso-PC to promote the formation of micellar forms.[280] In vitro studies revealed that in contrast to the reacylation mechanism, the transacylase reaction can utilize lyso derivatives of sn-1-phosphatidylcholine as well as the natural sn-3-phosphatidylcholine isomer.[281] Investigations in which both radioactive isomers were injected into adult rats failed to detect any radioactive conversion from the unnatural 3-palmitoyl-sn-1-phosphatidylcholine isomer.[281] This demonstrated that in vivo DPPC is only produced from lyso-PC through reacylation or some other stereospecific pathway. Further evidence against this mechanism includes the low activity of this enzyme relative to acyltransferase in type II cells from fetal and adult lung.[253,282-284] The level of transacylase activity in lung varies greatly in different species.[285] The low activity of this enzyme in fetal rat lung in late gestation, during the period in which the production of pulmonary surfactant is rapidly increasing, also argues against an important role in vivo.[286]

Although the deacylation-reacylation cycle (mechanism a), originally proposed by Hill and Lands[287] is clearly involved in the introduction of polyunsaturated fatty acids into phospholipids in a number of tissues, it is apparent that this mechanism could also account for introducing palmitate into the 2-position of DPPC.[9,107,244,245] Kinetic studies suggest that rat lung microsomes contain two distinct acyltransferases which exhibit specificity for either saturated or unsaturated fatty acids.[288] Although unsaturated acyl-CoAs show higher reactivity when added separately,[288-290] competition experiments reveal a marked selectivity towards palmitoyl-CoA.[289-290] The pulmonary acyl-CoA pool is characterized by high levels of palmitate and stearate and very little polyunsaturated acyl-CoAs when compared to liver.[290] The pulmonary acyl-CoA:1-acyl-lyso-PC acyltransferase system shows a higher selectivity for palmitate. The formation of DPPC in lung may be promoted by relatively high endogenous levels of 1-acyl-lyso-PC relative to 2-acyl-lyso-PC.[290]

A number of studies have shown that exogenous palmitate is recovered primarily from the 2-position of PC in whole lung or isolated type II cells.[291-295] Some authors have reported that palmitate synthesized *de novo* from acetate or glucose is preferentially associated with the 1-position of disaturated PC.[296-298] The manner in which this metabolic selectivity is imposed has not yet been explained.

Early studies by Kyei-Aboage et al.[299] suggested that in addition to the reacylation (a) and transacylation (b) mechanisms, disaturated PC could be formed in vivo through an exchange reaction which did not require the addition of ATP plus CoASH (mechanism c). Evidence for such a mechanism was reported by Engle et al.[298] More recently, Nijssen and van den Bosch[300] have reported a cytosol-dependent remodeling of the PCs in rat lung microsomes, which leads to an increase in the level of disaturated species. Further investigations[301] revealed that rat lung microsomes catalyzed a CoA-dependent but ATP-independent exchange of the fatty acids from the 2-position of membrane-bound PCs to the 2-position of certain lysophosphatides. This exchange mechanism, which preferentially utilized PCs containing arachidonate or linoleate, transferred these fatty acids to either lyso-PE, lyso-PS, or lyso-PG but not to lyso-PA. Using a similar approach, Stymne and Stobart[302] observed a reaction in rat lung microsomes by which fatty acids at the 2-position of PC are exchanged with fatty acyl-CoA through a reversal of acyl-CoA:1-acyl-lyso-PC acyltransferase. In the presence of CoA but without any acyl-CoA, lyso-PC accumulated. Incubations with mixed acyl-CoAs in ratios equivalent to the molar ratios of the fatty acids present at the 2-position of PC revealed a marked selectivity for the incorporation of palmitate into the diacyl lipid fraction. These studies suggest that 1-acyl-lyso-PC acyltransferase functions in both the forward and the backward directions. The backward reaction favored the formation of unsaturated acyl-CoAs, which can be selectively utilized in the acylation of lyso derivatives to form PE, PS, and PG. In contrast, palmitoyl-CoA is preferentially incorporated into disaturated PC.

Although the possibility that DPPC could be formed directly by the *de novo* pathway had been suggested, early observations indicating that cholinephosphotransferase discriminated against disaturated diacylglycerols led to the conclusion that the auxiliary pathways were primarily involved in the production of DPPC.[51,107,243,244] Re-examination of this question has revealed not only that disaturated phosphatidate and diacylglycerols can be generated by the *de novo* pathway but also has presented strong evidence indicating that they can be directly incorporated into disaturated PC.[291,303,304] Evidence for the view that DPPC can be generated by the *de novo* pathway has also been obtained through studies with type II cells.[293,295,305,306] Time course studies with whole lung and with freshly isolated type II cells reveal a graduated increase in the incorporation of palmitate into disaturated species of PA and diacylglycerol.[296,303] The labeling of PG followed that of PA.[295] Although the extent to which the Kennedy pathway contributes to the formation of disaturated PCs cannot be directly determined, estimates indicate that approximately 25% of the total PC is produced in this manner.[295]

C. Biosynthesis of the Acidic Phospholipids in Pulmonary Surfactant

As indicated earlier, PG is virtually undetectable in most mammalian tissues other than lung. The inverse relationship between the levels of PG and PI in surfactant noted in human amniotic fluid (Figure 9) and the reported correlation between the presence of PG in tracheal aspirates and improvements in the clinical situation[94,307] have stimulated considerable interest in these lipids in lung. However, although early reports suggested that surfactant containing PG was superior to surfactant containing PI, it has become apparent that the nature of the acidic phospholipid has little effect on the physical or biological properties.[308,309]

The pathway for the biosynthesis of the acidic phospholipids is initiated by the formation of the liponucleotide, CDP-diacylglycerol, from PA and CTP (reaction 1, Scheme 3). The enzyme responsible for this reaction has been detected in pulmonary microsomal and mitochondrial fractions. CDP-diacylglycerol does not accumulate in tissues but acts as a phosphatidyl donor for the formation of PI (reaction 2, Scheme 3) and phosphatidylglycerophosphate, which is subsequently hydrolyzed to form PG. The initial reaction in the formation of this latter surfactant lipid involves the transfer of the phosphatidyl

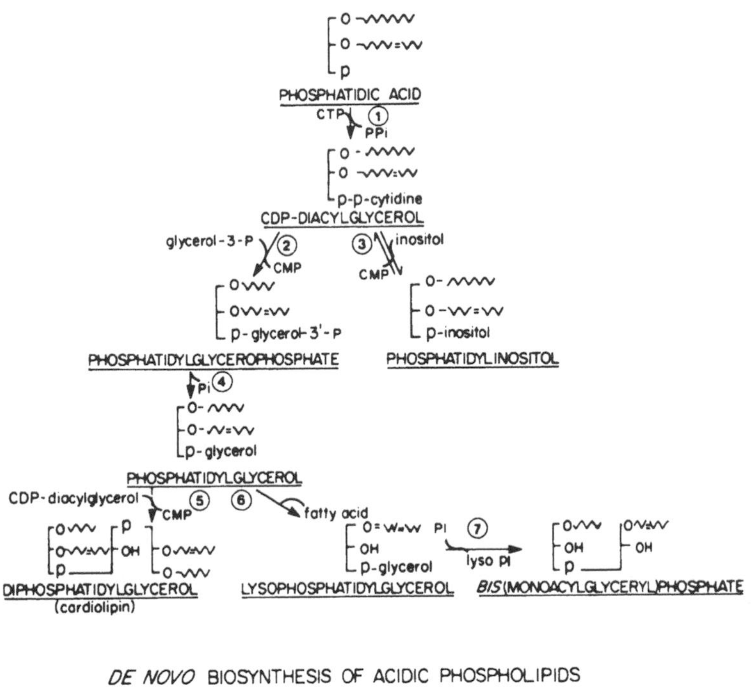

DE NOVO BIOSYNTHESIS OF ACIDIC PHOSPHOLIPIDS

Scheme 3.

moiety to the 1-hydroxyl of *sn*-glycerol-3-phosphate to yield *sn*-3-phosphatidyl-1'-*sn*-glycerol-3'-phosphate (reaction 2, Scheme 3). As indicated earlier, dihydroxyacetonephosphate will not replace *sn*-glycerol-3-phosphate in this reaction. The release of Pi from phosphatidylglycerophosphate is catalyzed by phosphatidylglycerophosphate phosphohydrolase, (reaction 4, Scheme 3), an enzyme which will be further discussed in a later section. Under the conditions utilized by most investigators, phosphatidylglycerophosphate does not accumulate so that reactions 2 and 3, Scheme 3 are assayed simultaneously.

Phosphatidate cytidylyltransferase, the enzyme responsible for the formation of CDP-diacylglycerol, has been studied in lung microsomal and mitochondrial fractions.[310,311] The formation of this liponucleotide constitutes the initial committed step for the formation of cardiolipin (diphosphatidylglycerol) and lyso-*bis*-phosphatidate as well as for PG and PI. Developmental studies reveal that in the rat and the rabbit, the specific activity of phosphatidate cytidylyltransferase increases in the microsomal fraction during the perinatal period. The mitochondrial activity remains relatively constant in the rabbit[310] and falls after birth in the rat.[312] With both species, the specific activity of the microsomes is lower than that of the mitochondria during the fetal period, but this situation is reversed in the adult. The apparent correlation between the specific activity of this enzyme in the microsomal fraction and surfactant appearance suggests but does not prove a functional role in surfactant synthesis.

PI is synthesized by CDP-diacylglycerol:inositol phosphatidyltransferase, an enzyme which is predominantly localized in the endoplasmic reticulum.[310,313] This enzyme reaction is reversible in lung as well as in other tissues.[313] The pH optimum for the forward direction was 8.8 to 9.4 while that for the backward reaction, which utilizes CMP, is 6.2. The specific activity of this enzyme increases in the microsomal fraction of rat and rabbit lung during the perinatal period and reaches its highest level in the adult.[310,312-314]

Microsomes from rabbit lung also possess an exchange reaction by which inositol can be incorporated into PI in the absence of exogenous CDP-diacylglycerol.[315] This exchange reaction is dependent upon CMP and is stimulated by Mn and Mg ions. Although this reaction could be explained by the production of CDP-diacylglycerol through the reversal of PI synthesis, followed by the resynthesis of this acidic lipid, the available evidence suggests that a separate process is involved.

Lung also contains a CMP-dependent incorporation of glycerol-3-phosphate into phosphatidylglycerophosphate and PG.[316] This pathway, which is inhibited by low concentrations of inositol, may involve the reversal of PI synthesis. This process, which is limited to the microsomal fraction, appears to use PI to generate PG.

Considerable effort has been expended in attempts to identify the subcellular source of the PG used for the formation of surfactant. In most tissues the only function of PG is to act as an intermediate in the biosynthesis of cardiolipin in the mitochondrion.[9] Although cardiolipin has not been extensively studied in lung, it is clear that pulmonary mitochondria can produce PG[310,312,317] and, presumably, at least part of this activity can be used for cardiolipin production. Microsomal synthesis of PG for surfactant would simplify matters, both in terms of the coordination of PI vs. PG production and in terms of a common system for transport to the lamellar bodies. Evidence for a microsomal contribution towards pulmonary synthesis was obtained with rat lung microsomes by Hallman and Gluck.[310,318] In addition, these authors have reported that in rabbit lung the specific activity of glycerophosphate phosphatidyltransferase increases threefold during the perinatal period.[310] The specific activity of this enzyme decreased in the mitochondrial fraction during development. It is important to note that the specific activity of the mitochondrial fraction remained higher than the microsomal fraction throughout development, with a fourfold excess being evident at birth.

In contrast, studies with either developing or adult rat lung demonstrated that virtually all of the glycerophosphate phosphatidyltransferase activity could be attributed to the mitochondria or to mitochondrial contamination of other fractions.[312,317] Fractionation of isolated type II cells from rat lung also revealed that only a small proportion of the total enzyme activity sedimented with the microsomal fractions.[319,320] However, examination of the properties of this enzyme revealed that, as had previously been reported for whole rabbit lung,[321,322] low concentrations of inositol specifically inhibited the microsomal enzyme.[320] These observations support a role for the endoplasmic reticulum in the production of PG and PI for surfactant. These studies are in good general agreement with other investigations, which showed that inositol starvation of Chinese hamster ovary cells produced a marked decline in PI and a concomitant increase in PG.[323]

XIII. MG^{2+}-INDEPENDENT PHOSPHATIDATE PHOSPHOHYDROLASE ACTIVITY IN LUNG

Phosphatidate phosphohydrolase was first described in plants by Kates in 1955.[324] The initial characterization of this enzyme in mammalian tissues was made in the early 1960s in brain,[325] intestine,[326] kidney,[327] liver,[328,329] and erythrocytes.[330] These investigations, which used aqueous emulsions of PA as the substrate, observed that the enzyme activity was primarily associated with the membranous fractions. Relatively high levels of phosphatidate were employed and with the exception of the erythrocyte, the effect of Mg^{2+} was not investigated. Interest in pulmonary phosphatidate phosphohydrolase and its properties was prompted by the finding that the specific activity of this enzyme increased in fetal rabbit lung[331] and that this activity could be found in human amniotic fluid.[332] Initial studies in lung used pH values between 5.5 and 7.0 and mM levels of PA. Most of the enzymatic activity was associated with the microsomal fraction, but activity was also found in the mitochondria and supernatant.[333-335] Low levels of Mg^{2+} had little effect but higher levels were inhibitory.[333-335] The release of Pi was also inhibited by F^-, Ca^{2+}, Mn^{2+} and Be^{2+}.[335,336] The activity was inhibited by the addition of lyso-PA or phosphatidylglycerophosphate but was only slightly affected by PC or by water-soluble substrates. Lyso-PA could act as a substrate to this enzyme.[334,335]

The presence of phosphatidate phosphohydrolase activity in human amniotic fluid and the observation that this activity paralleled the increase in L/S ratio[332,337,338] stimulated a detailed study of this enzyme in lamellar bodies and in isolated surfactant.[339-344] Phosphatidate phosphohydrolase in lamellar bodies from pig lung was less susceptible to inhibition by Mg^{2+} or Be^{2+} but more sensitive to heat inactivation than its microsomal counterpart.[336] The specific activity of this hydrolase in lamellar bodies was approximately double that in the microsomal fraction. Evidence for an independent phosphohydrolase activity in the lamellar bodies was also obtained through studies using the techniques of Dawson and McMurray,[345] in which the phosphohydrolase activity was determined in a series of tubes containing increasing amounts of microsomes but a fixed amount of lamellar bodies. Extrapolation of the phosphatidate phosphohydrolase activities to zero microsomal content revealed an activity associated with the lamellar bodies which could not be explained by microsomal contamination.[336]

The potential role of phosphatidate phosphohydrolase in the regulation of the production of diacylglycerol for surfactant PC prompted a comparative study on the properties of this hydrolase and phosphatidylglycerophosphate phosphohydrolase. The existence of a single enzymatic activity was inferred on the basis of similar inhibitory profiles for both activities with either mercury or heat-inactivation and from the observation that each substrate inhibited the degradation of the other phospholipid.[340] These studies led to the suggestion that an increase in pulmonary phosphatidate phosphohydrolase activity near term could account for the increased production of PG as well as PC (see Section XVII.B). The co-identity of phosphatidate phosphohydrolase and phosphatidylglycerophosphate phosphohydrolase was supported by studies by Benson[346] who observed that these two activities were present in highly purified surfactant from dog lung. Casola et al.[347] confirmed the presence of phosphatidate phosphohydrolase and phosphatidylglycerophosphate phosphohydrolase in rat lamellar bodies, but found differences in sulfhydryl sensitivity and in the effects of detergents and substrate specificities, which implied that these activities were not catalyzed by the same polypeptide.

The potential role for Mg^{2+}-independent phosphatidate phosphohydrolase in the control of the production of PC for surfactant was also supported by development studies in a number of different species. Initial studies by Schultz et al.[331] observed a 3.5-fold increase in phosphohydrolase activity in microsomes from fetal rabbit lung between 25 days gestation and term. The specific activity declined somewhat after birth but increased in the adult. The specific activity of the cytosolic enzyme was lower than that in the microsomes but increased eight- to tenfold between 23 and 25 days gestation and then remained relatively constant during prenatal and postnatal development. A doubling of the microsomal Mg^{2+}-independent phosphohydrolase activity was observed by Casola and Possmayer[348] between 25 days gestation and term. This corresponds to the period when choline incorporation into PC increased with slices of fetal lung.[349]

The specific activity of the Mg^{2+}-independent phosphatidate phosphohydrolase activity in rabbit fetal lung of 26 days gestation correlated with fetal cortisol levels.[137] Treatment with glucocorticoids led to an increase in this activity in fetal rabbit lung, as well as to an enhanced incorporation of choline into PC and precocious morphological changes.[350-353] On the other hand, induction of pulmonary maturation in this species by estradiol did not affect the Mg^{2+}-independent phosphohydrolase activity.[354,355] Treatment of pregnant rabbits with a triiodothyronine analog, 3,5-dimethyl-3'-isopropyl-L-thyronine, produced an increase in the Mg^{2+}-independent activity, but this required higher levels of analog than were needed to increase either choline incorporation into PC or PC content.[137]

Developmental studies with rat lung revealed little change in the Mg^{2+}-independent phosphohydrolase activity prior to birth but an increase in the specific activity of the whole homogenate and the microsomal fraction after delivery.[333,356-358] Little change was observed with the cytosolic activity. The specific activity of phosphatidate phosphohydrolase in homogenates of mouse lung increased markedly during the perinatal period. As indicated above, marked increases have also been observed in the activity of the Mg^{2+}-independent activity in human amniotic fluid. This increase parallels the elevation in the L/S ratio.[106,337,338] Increases in Mg^{2+}-independent phosphohydrolase activity have also been reported with explants of fetal rat, rabbit, and human lung.[116,241] Administration of hormones did not appear to affect the Mg^{2+}-phosphohydrolase activity in explants from rabbit or human lung.[116,241]

The subcellular distribution of phosphatidate phosphohydrolase has been examined in type II cells isolated from adult lung by Crecelius and Longmore.[359] Addition of magnesium had little effect on this activity. In contrast to whole lung, the greatest proportion of the Mg^{2+}-independent activity was located in the cytosolic fraction. Smaller amounts were found associated with the lamellar body and microsomal and mitochondrial fractions. Nonspecific phosphatase activity assayed with α-naphthyl phosphate was also found associated with the membranous fractions. The cytosolic activity hydrolyzed disaturated substrate slightly better than phosphatidate with an unsaturated fatty acid at the 2-position.

XIV. MEMBRANE-BOUND PHOSPHATIDATE PHOSPHOHYDROLASE

Early studies on phosphatidate phosphohydrolase in a number of tissues including liver,[328,329] brain,[325] intestine,[326] kidney,[327] and erythrocytes[330] demonstrated that most of the enzymatic activity is associated with the particulate fractions, with virtually no activity being present in the cell-free supernatant. Nevertheless, it was observed that despite this endogenous phosphatidate phosphohydrolase activity in microsomal and mitochondrial preparations, radioactivity from glycerol-3-phosphate accumulated in PA.[360-363] Further investigations indicated that the cytosolic fraction contained a factor which stimulated the production of triacylglycerol from phosphatidate.

These studies on the reconstitution of diacylglycerol and triacylglycerol synthesis using enzymatically generated membrane-bound intermediates suggested that cytosolic factors might contribute to triacylglycerol synthesis in vivo. In 1967, Johnston et al.[364] and Smith et al.[365] independently reported that the cytosolic factor was soluble phosphatidate phosphohydrolase which was active with membrane-bound phosphatidate but exhibited little activity with aqueous emulsions of this lipid. Further studies suggested that this activity required Mg^{2+}.[366] It is presently thought that the reason for the low activity on aqueous dispersions of PA observed with the cytosolic fraction was due to the presence of considerable amounts of calcium in PA prepared with phospholipase D.[249,367] The preparation of calcium-free PA requires stripping with Chelex-D or a similar reagent. It must also be emphasized that investigations of enzymatic activity utilizing lipid substrates are complicated by the insoluble nature of the substrate. Consequently, it is not always clear how the activities observed with aqueous dispersions of lipids are directly related to the reaction of interest in vivo.[368-370] This is particularly true of studies with membrane-associated enzymes and of situations in which it is necessary to use detergents. As will become clear, this caution applies to the present system. In order to avoid potential difficulties, lipid enzymologists developed methods for preparing labeled substrates endogenously generated on cellular membranes. The prevailing rationale was that such membrane-bound substrates would resemble more closely the biosynthetic intermediates and thereby yield information pertinent to the biosynthetic reaction under investigation. Observations made on a number of enzymes, including diacylglycerol acyltransferase,[371-373] cholinephosphotransferase and ethanolamine phosphotransferse,[371,373-375] phospholipase A_2,[273] and CTP:phosphatidate cytidylytransferase[376,377] using membrane-bound substrates, allowed a clearer interpretation of the specificities of these enzymes and their properties.

The initial studies on the pulmonary phosphatidate phosphohydrolase activities which use membrane-bound phosphatidate were conducted by Casola and Possmayer.[378] These authors used labeled phosphatidate endogenously generated on lung or liver microsomes from *rac*-glycerol-3-[^{32}P]-phosphate. The microsomes were collected by centrifugation and were heat inactivated before being used as substrate. The activity assayed in this manner has been designated as the membrane-bound phosphatidic acid phosphatidate phosphohydrolase. This nomenclature has proven confusing to some readers. Therefore, the enzyme (or enzymes) which utilizes membrane-bound phosphatidate as a substrate will be referred to as the membrane-bound PA phosphohydrolase in this review. To avoid confusion, the term membrane-associated will be used to indicate enzyme location.

It was observed that, in contrast to the Mg^{2+}-independent phosphatidate phosphohydrolase activity which was mainly associated with the membranous fractions, a large proportion of the membrane-bound phosphatidate phosphohydrolase was present in the cytosolic fraction. While the Mg^{2+}-independent phosphohydrolase showed pH optima of 6.0 to 6.5, the microsomal and cytosolic forms of the membrane-bound phosphatidate phosphohydrolase possessed pH optima of 7.0 and 7.5 to 8.0 respectively. The activity with the membrane-bound substrate was extremely sensitive to inhibition by Mn^{2+} or Ca^{2+}. Increasing the Mg^{2+} concentration produced a slight stimulation and then a gradual decrease in activity. EGTA produced an inhibition at higher concentrations.

The effect of EDTA was particularly interesting. Addition of this Mg^{2+}-chelator normally reduced the cytosolic activity by more than 80% (range 65 to 90%) and the microsomal activity by half. The residual activity in the presence of EDTA appears to be due to the Mg^{2+}-independent phosphohydrolase. Addition of increasing amounts of Mg^{2+} in the absence of EDTA led to a slight stimulation followed by a gradual inhibition. Addition of Mg^{2+} in the presence of EDTA led to a marked stimulation resulting in activities which were twofold greater than that observed in the absence of EDTA or $MgCl_2$ with the cytosol and 2.5-fold higher with the microsomes.

Studies on substrate specificity demonstrated that the addition of water-soluble substrates either had no effect or else mildly stimulated the membrane-bound phosphatidate phosphohydrolase activity. The exception was 3,4-dihydroxy-1-phosphonate, the phosphonate analog of glycerol-3-phosphate which produced a 30% decrease in activity. While addition of PC had little effect, the phosphohydrolase activity with membrane-bound substrate was depressed by low concentrations of lyso-PA, PG, or phosphatidylglycerophosphate.

Studies with a number of ionic and nonionic detergents revealed that Triton X-100 and Nonidet P-40 stimulated the membrane-bound phosphatidate phosphohydrolase in both cytosol and microsomes but inhibited the Mg^{2+}-independent activity.[379] Deoxycholate stimulated the activities observed with aqueously dispersed and membrane-bound substrate but only with the microsomal fraction.

These observations were consistent with the view that the phosphatidate phosphohydrolase activities in rat lung cytosol and microsomes, which utilized membrane-bound phosphatidate, could be distinguished from the Mg^{2+}-independent activities which hydrolyze aqueous dispersions of PA on the basis of pH optima, effects of ions, and the requirement for Mg^{2+}. This indicated that lung could, at least theoretically, contain two operationally distinguishable cytosolic phosphatidate phosphohydrolases and two operationally distinguishable membrane-associated activities. However, these results did not exclude the possibility that these apparent differences in the activity with membrane-bound and free PA could arise from alterations in the physical state of the substrate rather than the presence of independent enzymatic activities. Therefore, attempts were made to adopt experimental approaches which directly involved the enzymes rather than the physical state of the substrate.

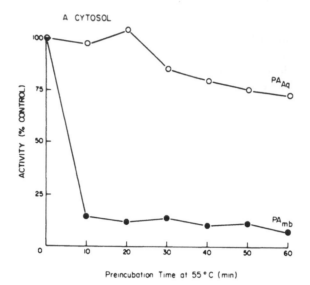

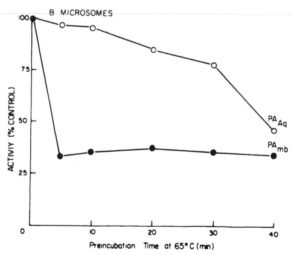

FIGURE 14.   Thermal stability profiles of the membrane bound phosphatidate and the Mg^{2+}-independent phosphohydrolase activities as a function of time at 65°C prior to assay. Abbreviations: PAAq, activity assayed with aqueously dispersed phosphatidate; PAMb, activity assayed with membrane-bound phosphatidate. (From Casola, P. G. and Possmayer, F., *Biochim. Biophys. Acta*, 664, 298, 1981. With permission.)

Preincubation of cytosol with low levels of trypsin or chymotrypsin prior to assay resulted in a marked stimulation of the membrane-bound phosphatidate phosphohydrolase activity while higher levels of protease resulted in a marked inhibition of this activity.[380] Protease treatment had little effect on the Mg^{2+}-independent activity in this fraction. The microsomal activity measured with membrane-bound substrate also proved more susceptible to protease treatment than the activity measured with the liposome substrate.

Similar results were observed in thermal stability studies, which revealed that over 75% of the membrane-bound phosphatidate phosphohydrolase activity in the cytoplasm and approximately 50% of the corresponding activity in the microsomes could be inactivated at 65°C with only a minor effect on the Mg^{2+}-independent activity (Figure 14). Taken together,

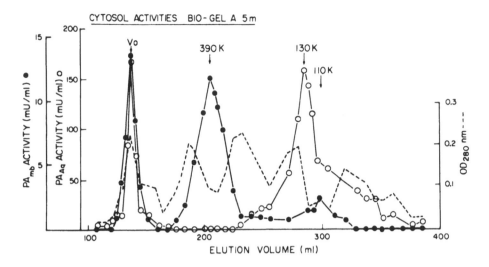

FIGURE 15.   Elution profiles for the membrane-bound (●) and Mg-independent (○) phosphatidic acid phosphohydrolase activities of rat lung cytosol on Bio-Gel A 5 m. Abbreviations as in Figure 14. (From Casola, P. G. and Possmayer, F., *Biochim. Biophys. Acta*, 664, 298, 1981. With permission.)

these results provided strong evidence indicating that separate enzymes were responsible for the degradation of membrane-bound and aqueous dispersions of PA. In addition, the similarity of the properties of the activity observed with membrane-bound substrate in microsomes and cytosol suggested a relationship between these two enzymes. The Mg^{2+}-independent activities in the cytosol and microsomes also exhibited similar properties.

Further evidence for independent phosphohydrolase activities acting on membrane-bound and aqueously dispersed PA came from studies[380] in which rat lung cytosols were precipitated with ammonium sulfate and fractionated on Bio-Gel A 5m columns (Figure 15). Under the standard assay conditions, 25 to 35% of both activities were eluted in the void volume. With membrane-bound PA as a substrate, a major peak (55% total) with an apparent $M_r =$ 390 kdaltons was observed followed by a minor peak (12% total) with and apparent $M_r =$ 110 kdaltons. When aqueously dispersed PA was used as the substrate, there was a major peak with an apparent $M_r = 130$ kdaltons followed by a broad shoulder which overlapped with the minor $M_r = 110$ kdalton peak observed with membrane-bound PA. The total peak accounted for 75% of the aqueously dispersed phosphatidate phosphohydrolase activity with approximately 25% being localized in the shoulder.

While the fractionation procedure afforded a separation of the cytosolic phosphohydrolase activities in reasonable yield (36% membrane-bound PA phosphohydrolase activity and 40% aqueous phosphatidate) the purification was low, only two- to threefold over the values observed in cytosol. The major difficulty encountered was that after ammonium sulfate fractionation the activity observed with membrane-bound PA declined by 50% overnight at 4, −20, or −80°C. Activities observed with aqueously dispersed PA were relatively more stable.

The effects of $MgCl_2$ and EDTA on the cytosolic fractions separated on Bio-Gel A-5m are illustrated in Figure 16 A and B. Addition of Mg^{2+} inhibited the hydrolysis of membrane-bound PA. EDTA virtually abolished the activity of the 390-kdalton fraction, but Mg^{2+} plus EDTA resulted in a modest stimulation over control values. The 130-kdalton fraction which hydrolyzes aqueously dispersed PA did not degrade the membrane-bound substrate under any of the conditions examined.

The most interesting effect noted with aqueously dispersed PA was that the 390-kdalton peak could hydrolyze this substrate in the presence of $MgCl_2$. The lack of any activity in

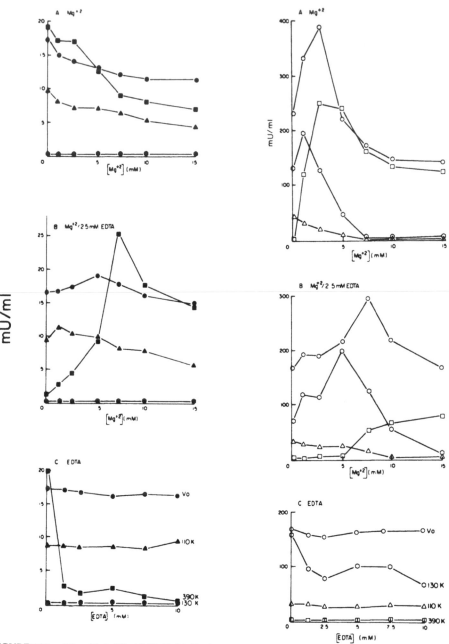

FIGURE 16A.   The effect of A, Mg^{2+}; B, Mg^{2+} plus EDTA; and C, EDTA on the membrane-bound PA phosphatidate phosphohydrolase activities from rat lung cytosol separated on Bio-Gel A 5 m. Two fractions were pooled from the V_0 peak (●), the 390-kdalton peak (■), the 130-kdalton peak (◉) and the 110-kdalton peak (▲); the PA_{Mb} phosphohydrolase activities were determined with the indicated addition of Mg^{2+} and/or EDTA. (From Casola, P. G. and Possmayer, F., *Biochim. Biophys. Acta*, 664, 298, 1981. With permission.)

FIGURE 16B.   The effect of A, Mg^{2+}; B, Mg^{2+} plus EDTA; and C, EDTA on the Mg^{2+}-independent phosphatidate phosphohydrolase activities separated from rat lung cytosol on Bio-Gel A 5 m. Two fractions were pooled from the V_0 peak (○), the 390-kdalton peak (□), the 130-kdalton peak (◯), and the 110-kdalton peak (△); the activities were determined with the indicated addition of Mg^{2+} and/or EDTA. (From Casola, P. G. and Possmayer, F., *Biochim. Biophys. Acta*, 664, 298, 1981. With permission.)

the absence of Mg^{2+} or in the presence of EDTA attests to the Mg^{2+} dependency of this fraction. This was the first indication of a pulmonary phosphatidate phosphohydrolase which exhibited an absolute dependency on Mg^{2+}. This also explains the absence of a 390-kdalton peak with aqueously dispersed substrate. Some Mg^{2+} dependency was observed with the void volume and with the 130-kdalton fraction. This latter fraction appears to require high levels of substrate for activity.

In order to obtain further insight into the potential relationship between the phosphohydrolase activities in the cytosol and those associated with the microsomal membranes, rat lung microsomes were partially solubilized with 0.5% Triton X-100/20% glycerol and precipitated with ammonium sulfate. The solubilized preparations demonstrated a 5.5-fold increase in the activity observed with membrane-bound PA but a slight depression of the activity assayed with free PA. Fractionation of the Triton X-100 treated cytosols on Bio-Gel A 1.5-m columns (Figure 17) revealed a small amount of phosphohydrolase activity toward both substrates, which eluted with the void volume. The microsomal elution pattern was characterized by a major peak which eluted at a slightly higher elution volume than the 390-kdalton activity in the absence of detergent. This peak contained a high proportion of the activity towards either membrane-bound or aqueously dispersed PA. A small activity peak which specifically hydrolyzed the aqueously dispersed phosphatidate eluted at a volume corresponding to the 130-kdalton peak in the absence of detergent. Purifications of approximately 15-fold were obtained for the activity directed toward membrane-bound substrate but only 3.5-fold with the activities observed with free PA.

For comparative purposes, rat lung cytosol was treated with Triton X-100, fractionated with ammonium sulfate, and eluted as above. The elution pattern for the detergent-treated cytosol was identical to that observed with the microsomal extracts except that the third peak which eluted in a position corresponding to the 110-kdalton cytosol peak in the absence of detergent possessed the largest proportion of the activity toward aqueously dispersed substrate. As in the case of the microsomal extract, this peak did not hydrolyze membrane-bound substrate. The purifications obtained with the cytosolic enzyme in the presence of Triton X-100 were similar to those observed with the microsomal extracts. These studies suggested a potential relation between part of the phosphohydrolase activities in the microsomes and the cytosol. Because the effects of detergents on protein configuration are variable, no conclusion could be made about the relationship between these activity peaks and the peaks eluted in the absence of detergent. Additional studies showed that Triton stimulated the cytosolic phosphohydrolase activities observed in the void volume but inhibited the Mg^{2+}-stimulation of the 390-kdalton cytosolic peak observed with aqueous dispersion of PA and abolished the corresponding activity with the 130-kdalton peak.

The fractionation studies described in this section proved useful in defining the Mg^{2+}-dependent and Mg^{2+}-independent phosphatidate phosphohydrolase activities of rat lung cytosol. Under the standard assay conditions, the Mg^{2+}-dependent phosphatidate phosphohydrolase activities conformed most closely to the activities observed with membrane-bound substrate, while the Mg^{2+}-independent phosphatidate phosphohydrolase activities conformed most closely to the hydrolase activities observed with aqueous PA (mainly represented by the 130-kdalton peak). Although the 110-kdalton appears to utilize substrate either with or without Mg^{2+}, developmental studies to be discussed later indicated that two separate activities are present. The observation that the 390-kdalton peak could be observed with aqueous dispersions of PA in the presence of $MgCl_2$ suggested that a major difference between the activities observed with the two substrates might be related to the presence of Mg^{2+} in the membrane-bound substrate. The radioactive membrane-bound substrate is prepared in the presence of Mg^{2+}, while the aqueously dispersed PA is converted to the sodium salt. Sturton and Brindley[381] also concluded that the presence of Mg^{2+} in PA prepared on microsomal membranes could explain some of the differences in the activity observed with

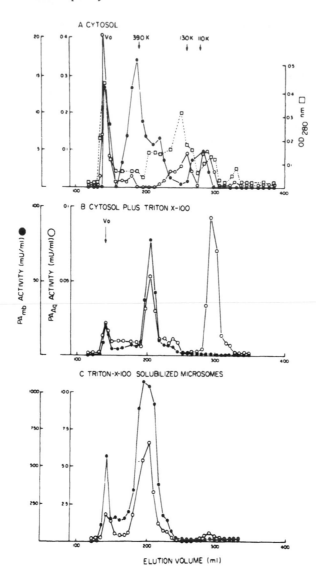

FIGURE 17. Comparison of the elution profiles of the membrane-bound (●) and Mg^{2+}-independent (○) phosphohydrolase activities from untreated and Triton X-100-treated cytosol and from Triton X-100-treated microsomes on Bio-Gel A 1.5 m. Abbreviations as in Figure 14. (From Casola, P. G. and Possmayer, F., *Biochim. Biophys. Acta*, 664, 298, 1981. With permission.)

this substrate and with aqueous dispersions of PA. It should, however, be emphasized that while the cytosolic activities observed with these two substrates show some relation, they are clearly distinguished by the 130-kdalton activity peak which was stimulated by Mg^{2+} and Mg^{2+} plus EDTA but did not hydrolyze membrane-bound PA.

A. Kinetic Properties

The realization that the presence of Mg^{2+} in membrane-bound PA could explain part of the difference between the activity observed with this substrate and with aqueously dispersed PA led to an investigation of the kinetic parameters observed with the phosphohydrolase activities in lung. In order to investigate the possibility that part of the anomaly observed

Table 3
KINETIC CONSTANTS FOR THE DIFFERENT PHOSPHATIDATE
PHOSPHOHYDROLASE ACTIVITIES ASSOCIATED WITH RAT LUNG
CYTOSOLIC AND MICROSOMAL FRACTIONS[379,387]

	Substrate							
	Membrane-bound phosphatidate		Lipid vesicle phosphatidate		Aqueously dispersed phosphatidate		PA:PC liposomes (Mg^{2+}-dependent)	
Fraction	K_m	V_{max}	K_m	V_{max}	K_m	V_{max}	K_m	V_{max}
Cytosol	14.3	110	10.5	100	250	2,200	215	6,800
Microsomes	14.3	400	9.7	256	750	70,000	55	1,600

Units: K_m, μmolar; V_{max}, μmol/min/mg protein.

Note: Lipid vesicle phosphatidate refers to the substrate obtained by sonicating the lipids extracted from ^{32}P-PA loaded microsomes.

between membrane-bound substrate and aqueously dispersed substrate could be related to the proteins remaining with the heat-denatured microsomes, the lipids were extracted from the labeled microsomes and sonicated in 0.9% NaCl. Since this sonicated substrate is clear, it presumably forms vesicles of some sort, probably liposomes, and has been designated lipid vesicle PA.

Although with either membrane-bound [^{32}P]PA, lipid vesicle [^{32}P]PA, or aqueously dispersed [^{32}P]PA, the situation can be described as heterogeneous catalysis,[368-370] plots of velocity vs. the concentration of membrane-bound or lipid vesicle [^{32}P]PA yielded reasonably hyperbolic curves which generated linear Lineweaver-Burke double reciprocal plots with both the cytosolic and microsomal fractions. In contrast, plots of the release of ^{32}Pi with increasing concentrations of aqueously dispersed PA did not yield Michaelis-Menten type kinetics, but were composed of sigmoidal curves which terminated in a velocity curve parallel to the abscissa. Similar results have been reported for rat lung microsomes by Ravinuthula et al.[333] and Mavis et al.[334]

Examination of the kinetic parameters revealed a close approximation for the apparent K_m and V_{max} values for the cytosol and microsomal fractions when membrane-bound PA and lipid vesicle PA were used as substrate. The apparent K_m and V_{max} values obtained with aqueously dispersed substrate were more than an order of magnitude greater than those observed with the other substrates (Table 3). Due to the aggregated nature of the substrate, the K_m values must be considered as effective rather than actual concentrations. These values will depend, to some extent, on the mode of substrate preparation. Nevertheless, the apparent K_m values observed with membrane-bound PA are similar to those previously reported for guinea pig brain cytosol.[382] In addition, the apparent K_m values for the aqueously dispersed substrate are similar to those previously reported for kidney microsomes,[327] brain microsomes,[325] and for cytosol and microsomes from adipose tissue.[383,384] The similarity between the kinetic parameters observed with membrane-bound and lipid vesicle PA indicated that the activities acting on these substrates in rat lung cytosol and microsomes may be related. These experiments also suggest but do not prove that the activities observed with aqueous phosphatidate are catalyzed by separate enzymatic entities.

B. Developmental Studies

As indicated earlier (Figure 8), in most species studied, the production of pulmonary

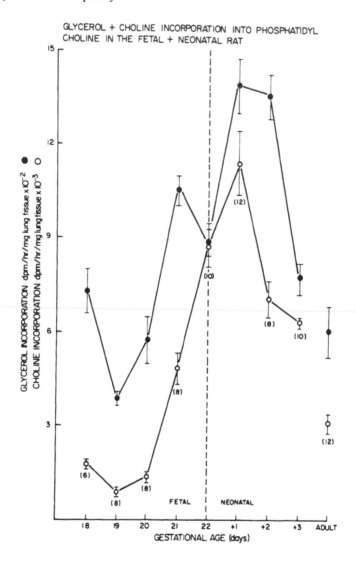

FIGURE 18.   Incorporation of radioactive glycerol and choline into lipids with slices of rat lung during development. (From Chan, F. et al., *Can. J. Biochem. Cell Biol.*, 61, 107, 1983. With permission.)

surfactant is initiated when gestation is approximately 80% complete. In keeping with this generalization, a marked increase in the incorporation of radioactive choline and glycerol into PC in rat lung slices occurs between day 20 of gestation and day 1 after birth, followed by a decline by day 3 (Figure 18).[286] The level of PC and disaturated PC in this species also increases between days 20 of gestation and day 1 after birth indicating that an increased production of PC is induced during this period and implying that at least part of this increased synthesis is related to surfactant production. The developmental profiles for the phosphatidate phosphohydrolase activities observed with membrane-bound and aqueously dispersed substrate are presented in Figure 19. There was little change in the specific activity of the Mg^{2+}-independent activity observed with aqueously dispersed PA in the whole homogenate, the microsomes, or the cytosol during the fetal period. An increase was observed in the specific activities of the microsomes and to a lesser extent of the homogenate following birth. Ravinuthula et al.[333] and Filler and Rhoades[358] also observed an increase in the specific activity of the Mg^{2+}-independent phosphatidate phosphohydrolase in the particulate fractions

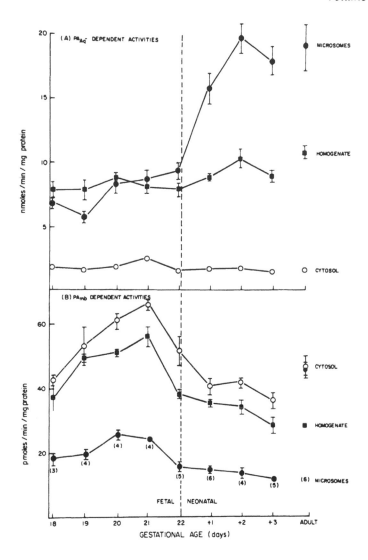

FIGURE 19. Developmental profiles of the phosphatidate phosphohydrolase activities in the fetal, neonatal, and adult rat lung. Phosphohydrolase activities were determined with (A) aqueously dispersed and (B) membrane-bound PA as substrates in whole lung homogenates (■), microsomes (●), and cytosol (○). Abbreviations: PA_{Aq}, aqueously dispersed phosphatidate; PA_{mb}, membrane-bound phosphatidate. (From Casola, P. G. and Possmayer, F., *Biochim. Biophys. Acta*, 665, 177, 1981. With permission.)

of rat lung at birth. It should be noted that this increase in specific activity occurred after the increase in the incorporation of radioactive precursors into PC and the elevation in PC content alluded to above (Figure 18). It was of interest to note that the specific activity of the microsomal fraction was similar to that observed for the whole homogenate during the fetal period but increased after birth.

In contrast to the Mg^{2+}-independent activities, the specific activities of the membrane-bound phosphatidate phosphohydrolase activity demonstrates a 1.5- to 1.6-fold increase between 18 and 21 days gestation. Virtually identical patterns were observed when lipid vesicles prepared from [^{32}P]-labeled microsomal lipid were used as the substrate. The developmental increase in specific activities preceded the augmented incorporation of radioactive precursors into PC by 1 to 2 days.

During these studies, it was observed that there was an overall increase in the protein content of the microsomal and cytosolic fractions per gram lung during development. The protein content of the microsomal fraction increased over twofold between day 18 and the first neonatal day. The cytosol protein content also increased by approximately 50% during the perinatal period. As a result of these alterations in protein content, a contrasting picture emerged when the phosphohydrolase activities were depicted as activity per gram lung. When expressed in this manner, increases in the Mg^{2+}-independent activities were also observed prior to birth. The elevations in the activities assayed using membrane-bound or lipid vesicle [^{32}P]PA were still apparent between 18 to 20 days gestation but the following decline noted with the specific activity profiles was much less evident.

The developmental patterns for the phosphohydrolase activities in developing rat lung were also examined by chromatography of cytosols in Bio-Gel A 5-m columns (Figure 20). Except for an increase in the relative proportion of activity eluting in the void volume, there was little change in the profile for the membrane-bound phosphatidate phosphohydrolase activity. The profile for the Mg^{2+}-independent activity observed with aqueously dispersed PA demonstrated a marked alteration during development. Although the phosphohydrolase activity measured with aqueously dispersed substrate was dominated by the 130-kdalton activity peak, virtually no activity was observed in this area with fetal and neonatal cytosols. In addition, although with the adult cytosols the shoulder of activity following the 130-kdalton peak could not be distinguished from the 110-kdalton activity observed with membrane-bound PA, the profiles obtained with the fetal and neonatal cytosols clearly showed that the Mg^{2+}-independent activity eluted later than the membrane-bound phosphatidate phosphohydrolase.

These developmental studies revealed that comparisons of the developmental patterns of the phosphatidate phosphohydrolase activities of rat lung are dependent not only on the form of the phosphatidic acid being assayed but also on whether the activity is expressed as specific or total activity. The marked increases in the total and specific activity of the membrane-bound phosphatidate phosphohydrolase occurred prior to the major elevation in the PC level and the augmented incorporation of radioactive precursors into this lipid. This time course is more consistent with a permissive rather than a regulatory role in PC bio-synthesis and surfactant production. The total Mg^{2+}-independent activity observed with aqueously dispersed PA increased somewhat in the microsomal fractions during the fetal period but even more after birth. This increase appears to parallel the pattern for the secretion of surfactant into the alveoli rather than to provide a means for increasing the production of surfactant PC.

The differences in the developmental patterns for the membrane-bound phosphatidate phosphohydrolase and the Mg^{2+}-independent activities support the presence of individual enzymatic activities in rat lung. This view was corroborated by the marked differences in the activity patterns observed after gel chromatography, in particular by the presence of the 130-kdalton Mg^{2+}-independent activity in the adult which was not observed during the perinatal period.

Similar overall conclusions were derived from studies conducted with fetal and adult rabbit lung.[348] In this species the incorporation of radioactive choline into PC with slices of fetal lung increased markedly between 27 and 29 days gestation resulting in a maximum at day 30 (term 31 days).[349] A concomitant surge in the tissue content of PC and disaturated PC occurs during this period.[349] The levels of PC in fetal lung lavage also increase during this period but a greater increase was observed after birth.[385,386] The specific and total activities of the membrane-bound phosphatidate phosphohydrolase increased in the microsomes and to a lesser extent in the whole homogenate between 26 days gestation and term but declined in the adult. The activities of the phosphohydrolase activity assayed with aqueously dispersed PA increased in the microsomal fraction to an even greater extent than did the activities

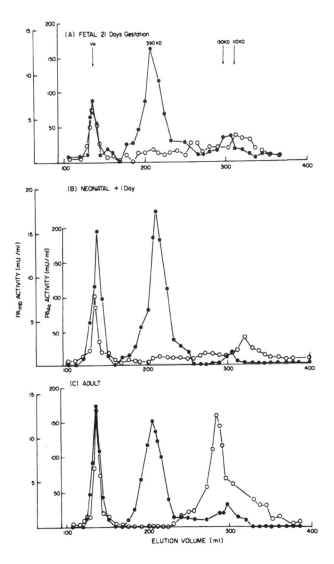

FIGURE 20. Elution profiles of the Mg^{2+}-independent ($\bigcirc$) and the membrane-bound phosphatidate ($\bullet$) phosphohydrolase activities from (A) 21-day fetal cytosols, (B) +1-day neonatal cytolsols, and (C) adult cytosols on Bio-Gel A 5 m columns. Abbreviations: PA_{Aq}, activity assayed with aqueously dispersed phosphatidate; PA_{mb}, activity assayed with membrane-bound phosphatidate.[357]

observed with the membrane-bound substrate. There was little change in the cytosolic activities assayed with either substrate between 25 days gestation and term. Both activities increased in the adult.

Fractionation of cytosol from adult rabbit lung on Bio-Gel A 5-m revealed membrane-bound phosphatidate phosphohydrolase activities in the void volume (15% total), a major peak with an apparent M_r = 390 kdaltons (44%), and minor peaks with M_r = 240 kdaltons (16%), and M_r = 110 kdaltons (25% total). Except for the presence of the minor activity at 240 kdaltons, this pattern resembled that observed with adult rat lung. Little change was observed during development. The fractionation pattern of the phosphohydrolase activity observed with aqueously dispersed phosphatidate also resembled the pattern for adult rat lung cytosol. A relatively large proportion of this activity (50%) eluted with the void volume, followed by a peak with an apparent M_r = 150 kdaltons (25% total), and a larger peak with M_r = 110 kdaltons (50% total). The 150-kdalton peak was not detected with fetal cytosols.

The specific activities of the phosphatidate phosphohydrolase activities observed with membrane-bound and aqueously dispersed PA increased in the microsomal fraction between 25 to 30 days gestation.[348] This corresponds to the fetal period when PC accumulates in this species. Schultz et al.[331] observed an increase in the cytosolic activities assayed with aqueously dispersed substrate between 23 to 25 days gestation. In contrast to the rat, little change in the protein content of the microsomal or cytosolic fraction was observed during fetal development. These results were consistent with a potential role for either the membrane-bound phosphatidate phosphohydrolase or the Mg^{2+}-independent phosphohydrolase activities associated with the microsomal fraction in the increased production of PC prior to birth. However, when taken together with the developmental profiles observed for developing rat lung, no consistent picture emerged to implicate any particular phosphatidate phosphohydrolase activity in either the microsomal or cytosolic fractions with the production of PC for surfactant.

XV. THE MG^{2+}-DEPENDENT PHOSPHATIDATE PHOSPHOHYDROLASE

The presence of Mg^{2+}-dependent, or at least Mg^{2+}-stimulated, phosphatidate phosphohydrolase activities in the microsomal and cytosolic fractions from liver and adipose tissue has long been recognized.[249,367] Early attempts to demonstrate Mg^{2+}-dependent phosphatidate phosphohydrolase activities in subcellular fractions from lung were uniformly unsuccessful.[333-335,378] The studies on the phosphohydrolase activities eluted from Bio-Gel A 5-m columns, described in the previous section, not only confirmed the presence of Mg^{2+}-dependent activities in pulmonary cytosol but also led to the realization that a basic difference between the phosphohydrolase activities observed with membrane-bound and aqueously dispersed PA arose from the presence of Mg^{2+} during the preparation of the former substrate. The similarity between the properties and the developmental patterns observed with membrane-bound PA and with sonicated vesicles prepared from the lipids extracted from [^{32}P]PA-loaded microsomes indicated that neither the proteins remaining nor the microsomal structure was essential for enzymatic activity. These considerations prompted a re-examination of the Mg^{2+} dependency of the phosphohydrolase activities in rat lung microsomal and cytosolic fractions. It was evident that in addition to the convenience of preparation, a chemically defined substrate would facilitate examination of kinetic properties and eliminate potential problems related to undefined amounts of other lipids and of other substrates.

The Mg^{2+}-dependent phosphohydrolase activities were assayed using mixed-lipid vesicles of equimolar egg PA:DPPC formed by sonication in 0.9% NaCl.[387] Maximal activities were observed with 0.5 mM equimolar PA:DPPC in the microsomes and 0.2 mM equimolar PA:DPPC in the cytosol. The incubation media contained 50 mM HEPES (pH 7.4), 1.25 mM EDTA, and 70 µg microsomal or cytosolic protein in a volume of 0.1 mℓ. Assays were performed in the presence or absence of 3.25 mM $MgCl_2$ resulting in a free Mg^{2+} concentration of 2.0 mM. Under these conditions, addition of Mg^{2+} produces a twofold increase in the release of ^{32}Pi with the microsomal fraction and a four- to fivefold increase in the release of radioactivity with the cytosolic fraction.

These studies employed [^{32}P]PA prepared by incubating diolein with [^{32}P]ATP in the presence of diacylglycerol kinase from *Escherichia coli*.[388] Previous studies have suggested that pulmonary phosphatidate phosphohydrolase is quite specific for PA.[336,340,378] Nevertheless, Sturton et al.[389,390] have observed that with rat liver microsomes, ^{32}Pi can be released from radioactive PA by a combination of phospholipase A_1 and A_2 activities followed by degradation of the glycerol-3-[^{32}P]phosphate formed to yield glycerol plus labeled inorganic phosphate.[249,367] In order to ensure that a similar activity was not interfering with studies on the phosphohydrolase activities in lung, comparative studies using the identical microsomes and cytosols were conducted using [^{32}P]PA and [^{14}C]PA prepared by incubating lung microsomes with [^{14}C]glycerol-3-phosphate.[452] The microsomal incubations showed that the Mg^{2+}-dependent release of radioactive diacylglycerol accounted for 94% of the radioactivity from [^{32}P]PA while the label in monoacylglycerol accounted for 7%. Approximately 7% of the total [^{14}C]label was associated with lyso-PA, glycerol-3-phosphate, and glycerol combined. These results indicated that little Mg^{2+}-dependent phospholipase A_2 could be present in rat lung microsomes. Extractions with acid molybdate suggested that 90% of the radioactivity released from the [^{32}P]PA was due to ^{32}Pi.

In contrast to the Mg^{2+}-dependent activities, 30% of the Mg^{2+}-independent radioactivity released was localized in [^{14}C]glycerol. This could arise either through phospholipase activity, as indicated above, or through degradation of diacylglycerol and monoacylglycerol by lipase activity.[291,375] Since addition of nonradioactive glycerol-3-phosphate did not affect the accumulation of ^{32}Pi, it appears that the latter route predominates.

Incubations with the cytosolic fraction revealed that the presence of Mg^{2+} produced a 16% greater release of ^{14}C-diacylglycerol than ^{32}P radioactivity. Small amounts of label were associated with lyso-PA, glycerol-3-phosphate, or glycerol, but the amounts due to Mg^{2+} accounted for less than 5% of total.

These results suggested that as in the case of the microsomes, the Mg^{2+}-dependent release of radioactivity from [^{32}P]PA was primarily related to phosphatidate phosphohydrolase rather than to other activities. However, these studies should be repeated using [^{32}P] and [^{14}C]-labeled substrates prepared in the same manner. Studies with [^{14}C]PA on PA-loaded microsomal membranes also demonstrated that [^{14}C]diacylglycerol was the major product and that the radioactivity associated with this neutral lipid was slightly higher than that released from the membrane-bound [^{32}P]PA.[453]

Velocity vs. phosphatidate concentration curves showed that under optimal levels of this divalent cation, the Mg^{2+}-dependent activities produced approximately hyperbolic curves (Figure 21). The activity was inhibited at higher levels of phosphatidate. Double reciprocal plots yielded apparent kinetic constants of $K_m = 55$ μM, $V_{max} = 1.6$ nmol/min/mg protein for the microsomal fraction. This is approximately fourfold higher than the values observed with membrane-bound PA, a result which may be explained by the low amounts of PA on loaded microsomes. However, the cytosolic studies yielded a surprisingly high $K_m = 215$ μM and a $V_{max} = 6.8$ nmol/min/mg protein which is similar to that previously observed for the activity with aqueously dispersed substrate (Table 3). No explanation can be given for the discrepancy between the Mg^{2+}-dependent activity in the cytosol and that observed with membrane-bound substrate.

The pH profiles of the Mg^{2+}-dependent phosphohydrolase activity in the microsomal and cytosolic fractions exhibited optima of 7.2 and 7.5 to 8.0 which were identical to those previously reported for membrane-bound substrate. Examination of the subcellular distribution of the Mg^{2+}-dependent activity demonstrated that as in the case of studies with membrane-bound substrate, the highest specific activity and over 90% of the total activity was associated with the cytosol. In addition, thermal inactivation studies revealed identical curves for the Mg^{2+}-dependent activity in the microsomes and cytosol, corresponding to those observed with the membrane-bound substrate. These investigations strongly indicate that the phosphohydrolase activity observed with membrane-bound PA is equivalent to the Mg^{2+}-dependent activity observed with equimolar PA:PC mixed lipid vesicles. The equivalence of these activities would also explain the effects of EDTA and of EDTA and $MgCl_2$ on the hydrolysis of the membrane-bound substrate.

The advantage of using a chemically defined substrate is illustrated through studies on the effect of Triton-X 100 on the phosphatidate phosphohydrolase activities of rat lung. Previous studies by Johnston's group on the Mg^{2+}-independent phosphatidate phosphohy-

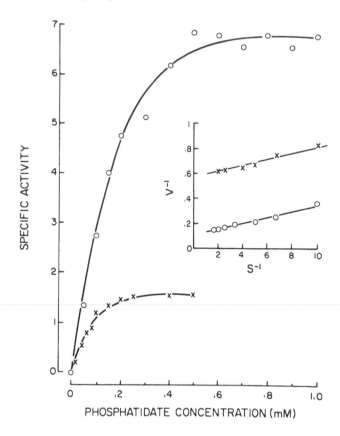

FIGURE 21.   Velocity vs. phosphatidate concentration curves and double reciprocal plots (insert) for the Mg^{2+}-dependent phosphatidate phosphohydrolase activities in rat lung microsomes ($\times$) and cytosol ($\bigcirc$). (From Walton, P. A. and Possmayer, F., *Anal. Biochem.*, 151, 479, 1985. With permission.)

drolase activities in pig lung lamellar bodies[340] and human lung explants[344,391] have utilized Triton X-100 to disperse the PA. Earlier studies from our laboratory showed that Triton X-100 produced a stimulation of the membrane-bound phosphatidate phosphohydrolase activities of rat lung microsomes and cytosol but an inhibition of activities observed with aqueously dispersed substrate.[379] In contrast, increasing amounts of this detergent led to a progressive inhibition of the Mg^{2+}-dependent activity observed with mixed PA:PC liposomes. An increase in the degradation observed in the absence of Mg^{2+}-dependent phosphatidate phosphohydrolase could be explained if the detergent produced a micellar dispersion which could be utilized by the normally Mg^{2+}-dependent enzyme in the absence of its Mg^{2+}-cofactor. This possibility constituted an important consideration toward defining the Mg^{2+}-dependent enzyme in relation to its Mg^{2+}-independent counterpart.

These experiments utilized the observations that pretreatment of microsomes and cytosol at 55°C for 15 min abolished the Mg^{2+}-dependent activity but led to only a small decrease in the Mg^{2+}-independent activities. Addition of 0.5% Triton X-100 abolished the Mg^{2+}-dependent activity but stimulated the Mg^{2+}-independent activity with the microsomes. Since the stimulation was also observed with microsomes preincubated at 55°C, it is apparent that the Mg^{2+}-independent activity is being selectively stimulated. Further studies revealed that Triton X-100 lowered both the K_m and the V_{max} of th Mg^{2+}-independent activity.

Addition of 0.5% Triton also led to an inhibition of the Mg^{2+}-dependent activity in the cytosol. However, in contrast to the microsomes, no stimulation was observed with the

Mg^{2+}-independent activity either before or after heat treatment. The studies with both fractions indicated that the addition of Triton stimulated the Mg^{2+}-independent activity at low levels of PA while inhibiting it at higher levels but obliterated the Mg^{2+}-dependent activity. The results also explained some of the anomalies arising from observations with aqueously dispersed and membrane-bound substrate.[379] Since it has been proposed that the activity observed with membrane-bound substrate is equivalent to the Mg^{2+}-dependent activity and that this particular activity is specifically involved in pulmonary glycerolipid metabolism this distinction is of practical and theoretical significance.

The effect of chlorpromazine on the phosphatidate phosphohydrolase activities in rat lung microsomes and cytosol was clarified using a similar approach. Although chlorpromazine acts as a competitive inhibitor of phosphatidate phosphohydrolase, it has been shown to stimulate the activity in the absence of Mg^{2+}.[367,390,392] This stimulation is masked in the presence of Mg^{2+}. As in the case of Triton X-100, these observations have important implications towards the identification of the Mg^{2+}-dependent phosphatidate phosphohydrolase activity. Studies with rat lung microsomes demonstrated that 1.0 mM chlorpromazine stimulated the total phosphatidate phosphohydrolase activity but reduced the Mg^{2+}-dependent activity somewhat. However, since the chlorpromazine stimulation was still present with heat-inactivated microsomes, it is evident that this stimulation cannot be explained through a replacement of Mg^{2+} by the cationic amphiphilic drug.[452]

Addition of chlorpromazine resulted in a marked inhibition of the Mg^{2+}-dependent phosphatidate phosphohydrolase activity in rat lung cytosol. In contrast to the results obtained with the microsomal fraction, this drug did not stimulate the Mg^{2+}-independent activity with either control or heat-treated cytosol. These studies supported the view that chlorpromazine did not replace Mg^{2+} as a cofactor for the Mg^{2+}-dependent activity in rat lung microsomes or in cytosolic fractions.

These investigations with Triton X-100 and chlorpromazine demonstrated that these agents could stimulate the Mg^{2+}-independent activities in the microsomes while not affecting the corresponding activity in the cytosol. Previous observations from this laboratory using aqueously dispersed PA had led to the conclusion that the Mg^{2+}-independent activities in the microsomes and cytosol were very similar if not identical. The basis for the difference in these results is not understood but could arise for two reasons. First, for more recent experiments, preparation of the cytosolic fraction includes perfusing the rat lungs with cold saline to remove all blood before homogenization. In addition, to ensure the absence of membranous contamination, only a portion of the clear supernatant fraction is withdrawn from the centrifuge tubes, rather than being decanted as in the former studies. Secondly, the Mg^{2+}-independent activity is now measured with the PA:PC liposomal substrate in the absence of Mg^{2+} rather than using the aqueously dispersed PA as in the previous format. Whether either of these alterations explains the difference in results is unknown and will require further study.

XVI. THE ROLE OF THE Mg^{2+}-DEPENDENT AND Mg^{2+}-INDEPENDENT PHOSPHOHYDROLASES IN THE PRODUCTION OF NEUTRAL GLYCERIDES AND PC IN LUNG

A number of investigators have demonstrated that it is the cytosolic activities measured with membrane-bound phosphatidate[365,366,372,393,394] or the Mg^{2+}-dependent or Mg^{2+}-stimulated activities[383,392,395-400] in liver and adipose tissue which show the greatest response to environmental stimuli and therefore could be involved in controlling the flow of substrates into triacylglycerol. More recently, it has become clear that translocation of the Mg^{2+}-stimulated cytoplasmic enzyme to the endoplasmic reticulum constitutes a major control system for production of neutral glycerolipids.[367] Studies conducted in lung have provided evidence for separate Mg^{2+}-dependent and Mg^{2+}-independent phosphohydrolase activities. However, although in retrospect, an argument could be made for the membrane-bound phosphatidate phosphohydrolase activities associated with the microsomes, there was no clear parallel between the developmental profiles for any of the four operationally distinguishable phosphatidate phosphohydrolases and the increased labeling of PC from radioactive choline or glycerol in both rat and rabbit lung. Studies on the induction of pulmonary maturation with glucocorticoids have identified an increase in the Mg^{2+}-independent phosphohydrolase in the microsomal fraction and this activity also increases in lung explants in culture.[116,241,359] The possibility exists that the increase in the membrane-associated Mg^{2+}-independent activity may be related to the production and accumulation of lamellar bodies. However, the enhanced incorporation of precursors into PC after maternal treatment with estradiol is not accompanied by an increase in any of the phosphatidate phosphohydrolase activities.

Initial attempts to identify the particular phosphatidate phosphohydrolase activity involved in the production of diacylglycerol for the formation of triacylglycerol, PE, and PC in lung were conducted by Casola and Possmayer[378,379] who examined the effect of cytosol addition on the disposition of radioactive PA in rat lung microsomes preloaded with [^{14}C] or [^{32}P]glycerol-3-phosphate. Addition of cytosol under conditions resembling those *in situ* resulted in a marked increase in the degradation of endogenously generated PA, especially during the initial period. Cytosol addition also led to an increase in the incorporation of radioactivity from [^{14}C]PA into PC in the presence of CDP-choline.[379] These studies suggested that the cytosolic activity could contribute to the formation of neutral lipids and PC in vivo but could not distinguish between the Mg^{2+}-dependent or independent activities.

Another approach to this problem took advantage of the observations of Germershausen et al.[401] who found that mouse microsomes washed with saline retained a higher proportion of the radioactivity incorporated from glycerol-3-phosphate at the PA level. Washing rat lung microsomes with increasing concentrations of NaCl led to a progressive increase in the proportion of radioactivity in PA from 63% to over 90% with a concomitant decrease in the labeling of diacylglycerol and triacylglycerol (Figure 22) from 37% to less than 10%.[402,403]

Salt washing reduced the total and the Mg^{2+}-dependent phosphatidate phosphohydrolase activity but had virtually no effect on the Mg^{2+}-independent activity (Figure 23). Most of the decrease in the Mg^{2+}-dependent phosphohydrolase activity occurred between 0.1 and 0.3 M NaCl. This corresponded to the salt concentration producing the maximal effect on the labeling of the neutral lipids (Figure 22).

These results were consistent with the view that the selective removal of the Mg^{2+}-dependent phosphatidate phosphohydrolase activity by salt-washing led to a diminution in the ability of the microsomes to produce diacylglycerol and therefore triacylglycerol. Other possibilities such as the removal of a factor which conferred Mg^{2+}-dependency on the Mg^{2+}-independent phosphohydrolase activity,[404] were considered. However, by restricting the volume of the wash, it was possible to demonstrate that the bulk of the Mg^{2+}-dependent activity removed from the microsomes could be recovered in the wash supernatant. Only a small proportion of the Mg^{2+}-independent activity was recovered in the washes (Figure 24).

Incubation of rat lung microsomes under the appropriate incubation conditions containing CDP-choline led to the labeling of PC from radioactive glycerol-3-phosphate. Washing the microsomes with 0.5 M NaCl resulted in an increase in the labeling of PA from 45 to over 80% (Figure 24) and a reduction in the labeling of monoacylglycerol, diacylglycerol, triacylglycerol, and PC. The ability of the salt-washed microsomes to metabolize PA was restored by addition of the wash supernatants. The ability of the washed microsomes to produce PC was also restored by the addition of rat lung cytosol or by the major Mg^{2+}-dependent phosphohydrolase activity eluted from Sephacryl S-400 gel columns. Addition of the Mg^{2+}-independent phosphohydrolase activity did not increase the proportion of radioactivity present

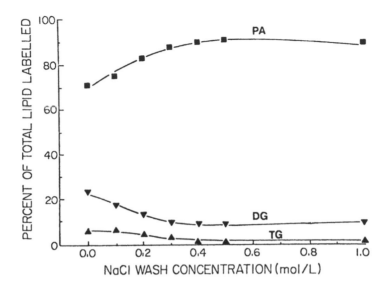

FIGURE 22. Effect of salt-washing on the incorporation of [¹⁴C]glycerol-3-phosphate into PA and neutral lipids; DG, diacylglycerol; TG, triacylglycerol. (From Possmayer, F. et al., in *The Physiologic Development of the Fetus and Newborn*, Jones, C. T. and Nathanielsz, P. W., Eds., Academic Press, New York, 1985, 235. With permission.)

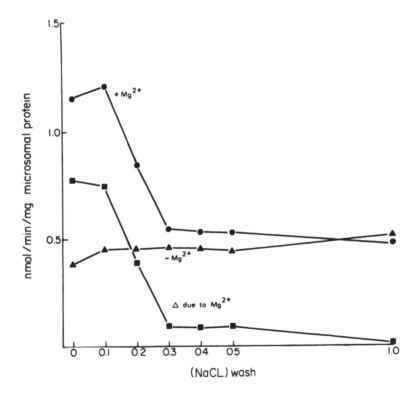

FIGURE 23. Effect of salt washing on the total Mg^{2+}-dependent and Mg^{2+}-independent phosphatidate phosphohydrolase activities of rat lung microsomes. (From Possmayer, F. et al., in *The Physiologic Development of the Fetus and Newborn*, Jones, C. T. and Nathanielsz, P. W., Eds., Academic Press, New York, 1985, 235. With permission.)

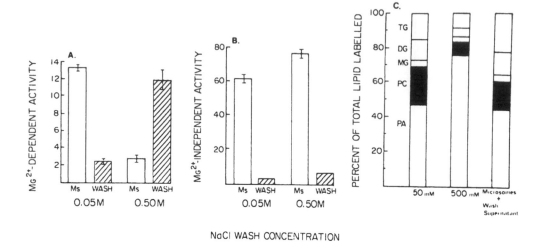

FIGURE 24.   Effect of salt-washing on (A) the Mg-dependent phosphohydrolase activity, (B) the Mg-independent phosphohydrolase activity, and (C) the incorporation of [^{14}C]glycerol-3-phosphate of rat lung microsomes. DG, diacylglycerol; TG, triacylglycerol; MS, microsomes; PC, phosphatidylcholine; PA, phosphatidate; MG, monoacylglycerol. (From Possmayer, F. et al., in *The Physiologic Development of the Fetus and Newborn*, Jones, C. T. and Nathanielsz, P. W., Eds., Academic Press, New York, 1985, 235. With permission.)

in neutral lipid or PC. These studies mark the most conclusive evidence implicating the Mg^{2+}-dependent phosphohydrolase activities in lung or in any tissue in the production of PC. Several additional factors must be considered. First, because of the strong diacylglycerol lipase activity associated with lung microsomes,[291,375] it was necessary to treat the microsomes with diisopropylfluorophosphate to inhibit this enzyme. Secondly, although approximately 90% of the Mg^{2+}-dependent phosphatidate phosphohydrolase activity detectable with added substrate was removed by salt washing, degradation of endogenously generated PA was only decreased by about half. Lastly, addition of cytosol, microsomal washings and, in particular, the later fractions from column chromatography produced an elevation in the incorporation of radioactive glycerol-3-phosphate into total lipids. While there is no reason to suspect that these factors interfered with the experimental results, unequivocal conclusions are not yet possible.

XVII. CONTROL OF PHOSPHOLIPID SYNTHESIS IN LUNG

Studies on the control of lipid metabolism have been hampered by the difficulties associated with the insoluble nature of many of the substrates and enzymes. Studies on the control of surfactant synthesis are further complicated by the fact that only part of the phospholipid being produced in lung is associated with surfactant and the type II cells producing this material account for only 10 to 15% of the total cellular complement of the lung. Despite these difficulties, experimental evidence has been obtained indicating that certain reactions play a role in the control of the production of PC and some of the other lipids present in surfactant. More recently, studies with isolated fetal and adult type II cells have extended our insight into these processes.

A. Control of PC Production

Initial attempts to identify the rate-limiting step in PC biosynthesis focused on cholinephosphotransferase (reaction 7, Scheme 1) the enzyme which catalyzes the terminal reaction in the production of this lipid. Farrell and Zachman[406] observed that direct injection of fetal rabbits *in utero* with 9-fluoroprednisolone led to a significant increase in the PC content of the lung, the incorporation of choline into PC with fetal slices, and the specific activity of cholinephosphotransferase. Treatment of decapitated fetal rats *in utero* with cortisol also promoted an increase in the specific activity of this enzyme.[407] The interpretation of this experiment is somewhat clouded because dipalmitoylglycerol was used as the phosphoryl-choline acceptor under conditions where this diacylglycerol is barely used by the enzyme.[9,264] Furthermore, it has become clear from more recent studies that treatment of pregnant rabbits with glucocorticoids and estradiol can lead to an increase in choline incorporation and PC content without significant alteration in cholinephosphotransferase activity.[349,350,352,354,355] This suggests that the induction of pulmonary maturation in this species can occur without an increase in the specific activity of this enzyme. The observation that laparotomy and fetal injection with saline can lead to an increase in cholinephosphotransferase activity in fetal rabbit lung offers a potential explanation for the previous findings.[408] Cholinephosphotrans-ferase activity is also augmented in this species by oxytocin-induced labor and increases after birth.[386] Whether these changes represent a direct or adaptive response to the increased demand for surfactant after birth remains unknown. Although the specific activity changes are not dramatic, increases in the total cholinephosphotransferase activity per gram lung, particularly in the microsomal fraction, are similar to the increases in the PC content of rat lung during the perinatal period.[312] Taken together, these results indicate that an increase in cholinephosphotransferase activity can accompany the increased production of PC for surfactant after birth, but does not appear to be necessary for an increase in PC formation during the fetal period.

Considerable interest has been generated by the finding that the specific activity of phos-phatidate phosphohydrolase increases several-fold in rabbit lung in late gestation[331] and in rat lung at birth.[333,357,358] The increase in the specific activity of the Mg^{2+}-independent activity is more prominent than are the alterations reported for any of the other enzymes involved in PC synthesis. It is clear that at least part of this increase could be related to the Mg^{2+}-independent activity associated with lamellar bodies and also with secreted surfactant. This has led to speculation concerning the role of lamellar bodies in surfactant produc-tion.[332,339-341] A potential role for this enzyme in regulation was further inferred by the suggestion that the enzyme responsible for the Mg^{2+}-independent hydrolysis of aqueously dispersed PA could also function in the degradation of phosphatidylglycerophosphate to produce PG.[340] This interesting hypothesis, which suggested that the control of both PC and PG for surfactant could be coordinated through a single regulatory enzyme, will be further discussed in the section on PI and PG. A considerable proportion of the Mg^{2+}-independent phosphohydrolase activity is localized in the microsomal fraction. The Mg^{2+}-independent enzyme activity does not appear to be concentrated in type II cells.[359,409] In addition, it has become clear that lamellar bodies do not contain the full complement of enzymes required for phospholipid synthesis.[9,425]

As elaborated in Section XVI, it is presently felt that the Mg^{2+}-dependent phosphatidate phosphohydrolase activity which can be measured either with membrane-bound substrate or PA:PC liposomes is involved in the production of diacylglycerol for PC synthesis. Studies in liver and adipose tissue indicate that it is the activity associated with or translocated to the microsomes which is important for glycerolipid metabolism.[249,367] Developmental studies with rat and rabbit lung do show an increase in the microsomal activity which parallels the increase in PC content in fetal rabbit and to a lesser extent in fetal rat lung. While these two observations are consistent with the view that the Mg^{2+}-dependent activity in the microsomes could play a regulatory role in surfactant production, it does not provide any convincing evidence.

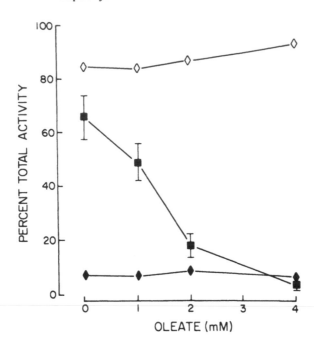

FIGURE 25.   Effect of oleate on the release of lactate dehydro-
genase ($\Diamond$), Mg^{2+}-dependent phosphatidate phosphohydrolase ($\blacksquare$),
and Mg^{2+}-independent phosphatase ($\blacklozenge$) from A549 cells permeated
with digitonin. (From Walton, P. A. and Possmayer, F., *J. Biochem.
Cell Biol.*, 64, 1135, 1986. With permission.)

The suggestion that phosphatidate phosphohydrolase could be involved in the control of
surfactant biosynthesis was supported by the observation that induction of pulmonary maturation in the rabbit with glucocorticoids produces a large, time-dependent increase in the
specific activity of this enzyme.[350,352] However, this increase was observed only with the
Mg^{2+}-independent activity. Furthermore, induction of pulmonary maturation with estradiol
did not produce an alteration in either Mg^{2+}-independent[354,355] or in the activity observed
with membrane-bound substrate.[355]

These observations suggest that an increase in the Mg^{2+}-dependent and, even more, the
Mg^{2+}-independent phosphatidate phosphohydrolase of lung can occur during development
and after the induction of pulmonary maturation. However, there is no direct evidence
implicating any of these changes with the overall control of PC production. The possibility
that, as in liver and adipose tissue, the Mg^{2+}-dependent activity could play a regulatory
role through translocation of cytosolic enzyme to the endoplasmic reticulum remains open.
Evidence for translocation has recently been obtained with A549 cells, a permanent cell line
derived from human lung which possesses type II cell characteristics. Exposing A549 cells
to free fatty acids (Figure 25) produced a rapid decrease in the proportion of the Mg^{2+}-
dependent activity which was released from digitonin-permeabilized cells.[410] Virtually none
of the Mg^{2+}-independent activity was released with fatty acid treatment. Incubation with
free fatty acids tends to increase the incorporation of choline into PC with these cells. Studies
in which increasing amounts of radioactive oleate were added to A549 cells revealed that
this fatty acid accumulated in PC, PE, and in monoacylglycerol, diacylglycerol, and triacylglycerol. With higher levels of free fatty acid, a larger proportion of the radioactivity
was localized in mono- and diacylglycerol. This pattern indicated that control of the production of PC and PE could be regulated by the production of the cytidine nucleotides.

Considerable evidence has accumulated which indicates that the production of PC can be controlled through the formation of CDP-choline.[411-413] Cholinephosphate cytidylyltransferase, the enzyme which catalyzes the reversible synthesis of this nucleotide, is found predominantly in the cytosol but significant amounts are also associated with the microsomal fraction. The specific activity of this enzyme shows little increase in cytosols from rabbit or rat fetal lung but increases after birth.[9,111,414-416] It has become evident that the postnatal increase in specific activity is related to enzyme activation rather than to enzyme production. Addition of total lung lipids, the acidic phospholipids PG or PI, or lyso-PE produces a marked increase in the specific activity of cytidylyltransferase with cytosols from fetal lung.[354,386,408,417-419] Very little effect is noted with cytosols from neonatal or adult lung or with the microsomal activity. Part of this activation can be ascribed to surfactant lipids but there is also an overall increase in nonsurfactant lipids in cytosol after birth.

Fractionation of cytosols on Sephadex columns revealed that in adult cytosol the enzyme is primarily present in a high molecular weight form eluting in the void volume, while the enzyme from fetal cytosols exhibits a molecular weight of 190 kdaltons. Addition of lipid to fetal cytosols induces the production of the high molecular weight form.[418] Thus, the lack of an activation effect with adult cytosols and presumably with the microsomes is related to the fact that the enzyme is already associated with lipid.

More recently, it has become evident that, as in the case of the Mg^{2+}-dependent phosphatidate phosphohydrolase, the activity of this enzyme in lung and in other tissues can be controlled through translocation from the cytosol to the endoplasmic reticulum.[411-413] Free fatty acids have also been implicated in this process.[419,420] The potential role of translocation in the control of PC synthesis in the neonatal lung will be discussed presently. The role of this in the regulation of PC in fetal lung will first be considered.

Evidence indicating that cholinephosphate cytidylyltransferase may catalyze a rate-limiting step in fetal lung arose from pool size measurements in rat and rabbit lung.[414,421,422] The changes in the pool sizes of choline and its intermediates in rat lung are depicted in Figure 26. Similar overall changes have been reported for rabbit lung during development.[421] In both species, choline levels tended to remain constant during the transitional period (rabbit, 25 to 28 days gestation; rat, 19 to 21 days) but increased slightly in the adult. The most notable alteration was a marked decrease in cholinephosphate during the transitional and mature periods and further after birth. In the rabbit, CDP-choline levels increased at the beginning of the transitional period but then fell as the lung achieved functional competence. This decline continued after birth.

In the rabbit these changes resulted in a decrease in the molar ratio of cholinephosphate to CDP-choline from 26:1 to less than 4:1. Pulmonary maturation in the rat is accompanied by a fourfold overall decrease in the level of cholinephosphate, a threefold decrease in CDP-choline levels, but a doubling of PC content. The changes in rabbit and rat lung have been interpreted as indicating that the cytidylyltransferase reaction is maintained at a pronounced disequilibrium in fetal lung due to the rapid utilization of CDP-choline. This disequilibrium is relaxed during development. These changes could also be interpreted by an increase in the level of CTP, but the level of this nucleotide falls in rabbit fetal lung during development.[423] The level of CMP showed a surprising increase during the perinatal period. These alterations are consistent with the view that pulmonary maturation is accompanied by a relative increase in the conversion of cholinephosphate through CDP-choline to PC.

Further evidence indicating that cytidylyltransferase catalyzes a rate-limiting reaction in fetal lung arose from studies on the induction of pulmonary maturation in rabbit fetuses.[355] Maternal treatment with estradiol on day 25 of gestation produced a significant increase in the incorporation of choline into PC with slices of fetal lung. Estradiol treatment produced a marked decrease in the pool size of cholinephosphate and a small but significant increase in the level of CDP-choline (Figure 27). Although the percentage increase in the pool size of PC was small, this resulted in an overall increase in choline content. Examination of the

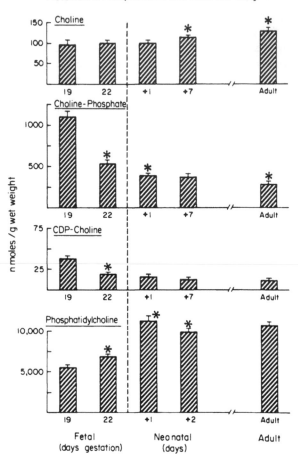

Pool Sizes of Phosphatidylcholine and its Choline-containing
Precursors in Fetal, Neonatal and Adult Rat Lung

FIGURE 26.   Pool sizes of choline and its derivatives in fetal, neonatal, and adult rat lung. (From Possmayer, F. et al., in *Novel Biochemical Pharmacological and Clinical Aspects of CDP-Choline*, Zappia, V. et al., Eds., Elsevier, Amsterdam, 1985, 91. With permission.)

activity of the enzymes in the choline pathway revealed a significant increase in cytosolic cytidylyltransferase activity but no change in the microsomal activity.[354,355,424] No changes were observed in the activity of choline kinase or cholinephosphotransferase. Addition of PG led to a smaller increase in the activity with treated than with control cytosols so that resulting activities were comparable. These results suggest that the increase in cytidylyltransferase activity is related to enzyme activation rather than to enzyme production. Although it has been postulated that this increase could be related to an increase in PG levels, detailed studies by Chu and Rooney[424,425] indicate that all of the cytosolic lipids may be involved in the activation of cytidylyltransferase. Interestingly, addition of PG to adult type II cells in culture results in an increase in cytosolic cytidylyltransferase activity which appears to be related to enzyme level.[409] Addition of total surfactant results in a decrease in enzyme activity.[426] This enzyme may also be controlled through phosphorylation:dephosphorylation.[427]

If cholinephosphate cytidylyltransferase catalyzed the only rate-limiting step in the biosynthesis of PC in lung, CDP-choline should not accumulate either during development or after hormonal induction (Figure 27). This suggests that a regulatory role could also be

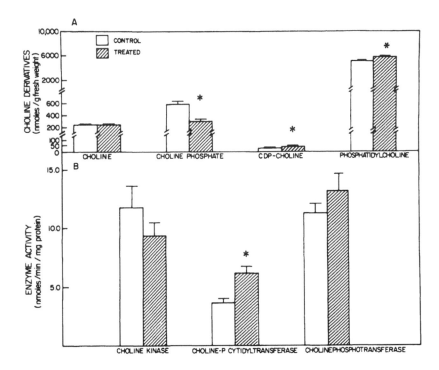

FIGURE 27.   Effect of maternal treatment with estradiol on the relative pool sizes and enzyme activities of the choline pathway for PC synthesis in fetal rabbit lung. Asterisks indicate a significant difference from control values. (From Possmayer, F., in *The Fetus and the Neonate*, Jones, C. T., Ed., Elsevier/North Holland, Amsterdam, 1982, 337. With permission.)

exerted at the level of cholinephosphotransferase. An increase in the reaction rate of the cholinephosphotransferase reaction would account for the fall in CDP-choline levels during development. A regulatory role for cholinephosphotransferase as well as for cytidylyltransferase has also been inferred from studies on the incorporation of choline into the intermediates of the choline pathway with slices of fetal rabbit lung.[421] Although this has not been examined in fetal rabbit lung, the levels of diacylglycerol increase in rat lung during development.[291,414,422]

Evidence indicating that cholinephosphate cytidylyltransferase catalyzes a rate-limiting step in PC synthesis during the neonatal period has been presented by Weinhold and his colleagues.[428,429] These authors demonstrated that allowing prematurely delivered rats to air-breathe for 3 hr resulted in a 30% increase in the incorporation of choline and ^{32}Pi into PC. Little effect was observed with other precursors of PC. This increased labeling of PC was attributable to a selective increase in the cytidylyltransferase activity associated with the microsomes. Air-breathing was also accompanied by a reduction in the pool size of cholinephosphate which accounted for the increased labeling of PC. The level of PC in the neonatal lungs was also increased. These authors suggested that free fatty acids, derived from enhanced pulmonary blood flow or from the degradation of serum triacylglycerols were responsible for the translocation of cytidylyltransferase from cytosol to endoplasmic reticulum. The activity of cytidylyltransferase in the microsomal fraction of neonatal liver also increases after delivery.[430]

Further evidence for a regulatory role for cholinephosphate cytidylyltransferase was obtained through pool size measurements on isolated type II cells. These measurements revealed that the levels of choline and its water soluble intermediates in adult type II cells were

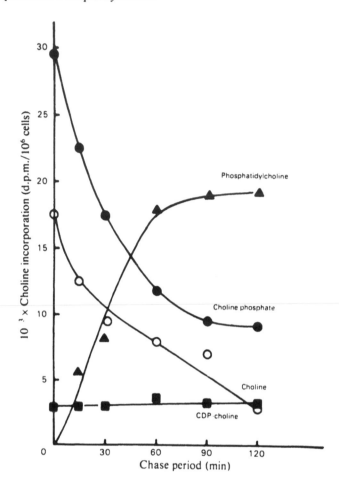

FIGURE 28.   Pulse-chase study on the metabolism of radioactive choline in type II cells isolated from 19-day fetal rat lung. (From Van Golde, L. M. G. et al., *Biochem. Soc. Trans.*, 13, 86, 1985. With permission.)

concentrated 25- to 100-fold compared to whole lung.[431] The levels of the cytidine nucleotides were also elevated. The cholinephosphate pool was considerably larger than the pools for choline and CDP-choline. These studies also indicated that the cytidylyltransferase reaction is maintained at a disequilibrium which regulates the flow of choline into PC. Pool size analysis studies with type II cells from fetal rat lung revealed similar levels of choline and cholinephosphate but lower levels of CDP-choline and PC.[432] The possibility that choline kinase can play a regulatory role was also considered.

The regulatory role of cytidylyltransferase was confirmed by pulse-chase experiments with fetal and adult type II cells.[431-434] The results obtained with fetal cells are given in Figure 28. Most of the radioactivity administered during the prelabeling period became associated with cholinephosphate. The pulse-chase showed that radioactivity from choline and cholinephosphate was rapidly metabolized to yield PC without accumulating in CDP-choline. Addition of FPF to fetal type II cells from rat lung produced a rapid increase in the labeling of PC from choline.[435] This increase was associated with an increase in cytidylyltransferase activity and a decrease in the pool size of cholinephosphate. Since the effect of FPF on choline incorporation attained a maximum within 60 min, it appears likely that this effect is also related to enzyme activation. Cholinephosphate cytidylyltransferase activity is also increased in choline-depleted cells.[426]

B. Control of the Production of PG and PI in Lung

As noted earlier, there is a striking alteration in the relative levels of PG and PI in surfactant from developing lung (see Figure 9). It is important to note that this change involves a marked increase in the levels of PG associated with surfactant in lamellar bodies and/or alveolar secretions. The absolute amounts of PI in surfactant do not decrease but actually continue to increase. The development stage for the reciprocal alterations in the acidic phospholipids varies with respect to birth in different species. Although there is considerable individual variation, the increase in the level of PG in amniotic fluid can be noted in normal human pregnancies around 36 weeks.[89,95,96] In the rat, this change occurs just prior to term,[436] whereas in the rabbit, the increased production of PG appears to coincide with birth.[310]

This alteration has attracted considerable scientific and clinical interest. The small amount of CDP-diacylglycerol present in lung indicates that the formation of this liponucleotide can be rate-limiting. The levels of both CDP-diacylglycerol:inositol phosphatidyltransferase and CDP-diacylglycerol:glycerophosphate phosphatidyltransferase tend to increase in microsomal preparations.[310,311] Consequently, since CDP-diacylglycerol acts as the phosphatidyl donor for both PI and phosphatidylglycerophosphate, and thus eventually PG, the most obvious explanation for the reciprocal relation between the acidic phospholipids would be a direct competition between free inositol and *sn*-glycerol-3-phosphate. Evidence for such a competition has been observed with lung microsomes,[321,322] with fetal lung explants,[149] with isolated type II cells,[437,438] and even with adult animals infused with inositol.[321] Fetal serum contains high levels of inositol.[9,439-441] This cyclitol appears to be required for rapid cell division, possibly because of its involvement in the PI cycle.[262]

It appears plausible that a major factor in the control of the formation of PG vs. PI in fetal lung involves the decline in the levels of inositol observed in human and fetal rat serum.[9,439-443] These decreases in inositol accelerate after birth. Alterations in inositol transport and metabolism may account for the delayed appearance of PG in amniotic fluid with diabetic pregnancies.[441] Studies with rabbit fetal lung revealed a decreased ability either to synthesize or to take up myoinositol as gestation advanced.[444] The fetal serum concentrations corresponded to the inositol levels at which inositol transport could be influenced.[440] However, despite these alterations, the levels of inositol in fetal rabbit lung were similar to those observed with adult lung and were greater than that required to inhibit the formation of phosphatidylglycerophosphate in vitro.[440,444] The reason why the total inositol content of the fetal tissues does not appear to reflect the availability of this sugar remains vague.[440,444]

Bleasdale et al. have proposed that in addition to the effects of inositol levels, the relative rates of the accumulation of the acidic phospholipids could be influenced by the levels of CMP in fetal lung.[241,423,439,440] This novel hypothesis is based on the observation that, although the reactions responsible for the synthesis of PI and phosphatidylglycerophosphate to PG are irreversible, the net effect is that elevated levels of CMP can promote the formation of CDP-diacylglycerol from PI but not from PG (Figure 29). Since PG and PI for surfactant can be produced by the endoplasmic reticulum (see Section XVII.C), the CDP-diacylglycerol originating from PI could be utilized for the synthesis of PI or PG. Consequently, although the production of PI for surfactant rises continuously during gestation, if CMP levels increased, an increasing proportion of PI could be recruited for the formation of PG.

CMP is released during the synthesis of PI and phosphatidylglycerolphosphate. It is also released from CDP-choline during the formation of PC. Quirk et al.[423] demonstrated changes in the levels of the cytidine nucleotides in rabbit lung during development which are consistent with a potential role for CMP in the increased formation of PG. These studies revealed that the relatively high levels of CTP decreased between 21 to 25 days gestation while the levels of CDP-choline increased. Between days 26 and 30 of gestation, there was a decline in the level of CDP-choline which was accompanied by a corresponding increase in CMP. This decrease in the fetal pulmonary CDP-choline content coincided with the elevation in the

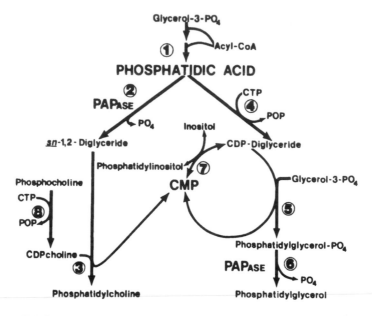

FIGURE 29.   The relationships of cytidine monophosphate (CMP) to phosphatidylcholine, phosphatidylinositol, and phosphatidylglycerol biosynthesis. The enzymes catalyzing the reactions are 1, acyl-transferases; 2, PAPase; 3, cholinephosphotransferase; 4, CTP:PA cytidylyltransferase; 5, CDP-diacylglycerol:glycerol-3-phosphate phosphatidyltransferase; 6, PAPase (PGPase); 7, CDP-diacylglycerol:inositol phosphatidyltransferase; 8, CTP:phosphocholine cytidylyltransferase. (From Bleasdale, J. E. and Johnston, J. M., in *Lung Development: Biological and Clinical Perspectives*, Vol. 1, Farrell, P. M., Ed., Academic Press, New York, 1982, 259. With permission.)

incorporation of choline into PC and the accumulation of this lipid. No increase was observed in the levels of the other nucleoside monophosphates, nor did the level of CMP increase in fetal liver.[423]

Bleasdale and Johnston[241,439,440] suggest that the increased utilization of CDP-choline for PC synthesis in dependent upon an elevation in the levels of diacylglycerol due to the increased phosphatidate phosphohydrolase activity observed in rabbit lung between 21 to 25 days gestation.[331] According to this theory, this is the ultimate source of the increased CMP levels. It may be recalled that Johnston et al.[340] have proposed that a single enzyme is responsible for the degradation of PA and phosphatidylglycerophosphate (see Section XIII). If this is the case, the increase in phosphatidate phosphohydrolase observed with rabbit fetal lung could promote the increase in both PC and PG noted in fetal rabbit lung near term.

Although the author does not feel that a single enzyme is responsible for the degradation of PA and phosphatidylglycerolphosphate and suspects that such an enzyme would be Mg^{2+}-dependent, he is intrigued by a number of the other observations which support this theory. The formation of PG from PI can be demonstrated in lung microsomes.[316] Addition of choline to permeabilized choline-depleted adult type II cells results in an incorporation of $[^{14}C]$-glycerol-3-phosphate into PG as well as an elevation in the formation of radioactive PC.[445] Likewise, addition of increasing amounts of CMP to permeabilized cells promotes a marked increase in the incorporation of choline into PC.[446] Addition of increasing amounts of CMP to permeabilized cells promotes a marked increase in the incorporation of $[^{14}C]$glycerol-3-phosphate into the diacylglycerol moiety of PG without affecting the labeling of PI. These observations indicate that the increase in the intracellular levels of CMP in fetal lung could promote the enhanced formation of PG.

In contrast, Batenburg et al.[437] have demonstrated that addition of choline to type II cells isolated from adult rat lung resulted in a simultaneous increase in the labeling of PG and PI from [^{14}C]glycerol-3-phosphate. Whether this difference in results is related to inositol levels is not clear. It is clear, however, that a number of factors are involved in the control of the production of the various lipids associated with pulmonary surfactant and considerable work is still required before a thorough understanding can be achieved.

XVIII. CONCLUSION

This review has attempted to summarize the present state of knowledge regarding the nature and composition of the surfactant system of the lung. The manner in which this material stabilized the alveoli through the production of a surface-active monolayer has also been discussed. As demonstrated through clinical trials with prematurely delivered humans, the presence of surfactant is particularly critical at birth when the newborn infant must clear its terminal airways of fetal pulmonary fluid and establish air-breathing.

Surfactant accumulates in the lung towards the end of gestation. The first association between pulmonary phosphatidate phosphohydrolase and surfactant arose from developmental studies which demonstrated an increase in the specific activity of this enzyme during the perinatal period. These early investigations, which emphasized the Mg^{2+}-independent activity, suggested a potential role for this activity in the control of the production of PC for surfactant. It has become clear that part of the Mg^{2+}-independent activity is associated with the lamellar bodies and is secreted into the alveoli with these organelles. Whether the Mg^{2+}-independent activity present in lamellar bodies represents the same enzyme as that associated with the microsomal fractions is not clear. Nor is it yet known whether this activity has a role either in the formation of the lamellar bodies or in the alveolus after secretion. The functional role of the Mg^{2+}-independent phosphatidate phosphohydrolase in the microsomal fraction also remains undetermined.

The Mg^{2+}-dependent activity present in lung cytosolic and microsomal fractions appears to be distinct from the Mg^{2+}-independent activity. The Mg^{2+}-dependent activity can be assayed either with membrane-bound PA or with PA:PC liposomes. While assays with these two substrates appear to measure the same enzyme, concern must be expressed about the differences in apparent K_m and V_{max}, particularly with the cytosolic fraction (Table 3). Whether the microsomal enzyme interacts with these substrates through membrane fusion must also be clarified.

It appears likely that the Mg^{2+}-dependent phosphatidate phosphohydrolase associated with the endoplasmic reticulum is involved in the production of PC surfactant. While not yet unequivocally proven, this enzyme appears to be identical to the Mg^{2+}-dependent phosphatidate phosphohydrolase present in the cytosol. It appears likely that glycerolipid metabolism in lung can be controlled through the translocation of Mg^{2+}-dependent phosphohydrolase from cytosol to the endoplasmic reticulum. Whether this translocation plays an important role in controlling the production of PC and DPPC for surfactant must still be determined.

ACKNOWLEDGMENTS

The author would like to express his gratitude to Dr. Paul Casola and Dr. Paul Walton for conducting many of the studies reported in this review. He would also like to thank Mrs. Fern MacDonald and Mrs. Jean Weick for editorial services, and Dr. Octavio Filgueiras, Mr. Mark Quirie, Mr. Jim Chung and Mr. Ramesh Izedian for reading the text. Supported by grants from the Medical Research Council of Canada.

REFERENCES

1. **Davson, H. and Danielli, J. F.,** *The Permeability of Maternal Membranes,* Cambridge University Press, London, 1943.
2. **Van Deenen, L. L. M.,** Phospholipids and biomembranes, *Prog. Chem. Fats Lipids,* 8, 1, 1965.
3. **Ansell, G. B., Hawthorne, J. N., and Dawson, R. M. C.,** *Form and Function of Phospholipids,* B.B.A. Library, Vol. 3, Elsevier, Amsterdam, 1973.
4. **Weissmann, G. and Claiborne, R.,** *Cell Membranes: Biochemistry, Cell Biology and Pathology,* HP Publ., New York, 1975.
5. **Snyder, F.,** *Lipid Metabolism in Mammals,* Vols. 1 and 2, Plenum Press, New York, 1977.
6. **Saunders, R. L.,** The chemical composition of the lung, in *Lung Development: Biological and Clinical Perspectives,* Vol. I, Farrell, P. M., Ed., Academic Press, New York, 1982, 179.
7. **Montfort, A., van Golde, L. M. G., and van Deenen, L. L. M.,** Molecular species of lecithins from various animal tissues, *Biochim. Biophys. Acta,* 231, 335, 1971.
8. **Holub, B. J. and Kuksis, A.,** Metabolism of molecular species of diacylglycerolipids, in *Advances in Lipid Research,* Paoletti, R. and Kritchevsky, D., Eds., Academic Press, New York, 1978, 1.
9. **Possmayer, F.,** Biochemistry of pulmonary surfactant during fetal development and in the perinatal period, in *Pulmonary Surfactant,* Robertson, B., Van Golde, L. M. G., and Batenburg, J. J., Eds., Elsevier, Amsterdam, 1984, 295.
10. **Thannhauser, S. J., Benotti, J., and Boncoddo, N. F.,** Isolation and properties of hydrolecithin (dipalmityl lecithin) from lung, its occurrence in the sphingomyelin fraction of animal tissues, *J. Biol. Chem.,* 166, 669, 1946.
11. **Avery, M. E.,** Lecture presented to the Canadian Society of Clinical Investigation, Montreal, Quebec, 1974.
12. **Comroe, J. H., Jr.,** in *Pulmonary and Respiratory Physiology,* Part 1, Comroe, J. H., Jr., Ed., Dowden, Hutchinson and Ross, Inc., Stroudsburg, Pa., 1976, 64, 212.
13. **Farrell, P. M.,** Overview of hyaline membrane disease, in *Lung Development: Biological and Clinical Perspectives,* Vol. 11, Farrell, P. M., Ed., Academic Press, New York, 1982, 23.
14. **von Neergaard, K.,** Neue Auffassungen uber einen Grundbegriff der Atemmmechanik. Die Retractionskraft der Lunge, abhangig von der Oberflachenspannung in den Alveolen, *Z. Ges. Exp. Med.,* 66, 373, 1929.
15. **Comroe, J. H., Jr.,** Premature science and immature lungs. I. Some premature discoveries, *Am. Rev. Respir. Dis.,* 116, 127, 1977.
16. **Comroe, J. H., Jr.,** Premature science and immature lungs. II. Chemical warfare and the newly born, *Am. Rev. Respir. Dis.,* 116, 311, 1977.
17. **Comroe, J. H., Jr.,** Premature science and immature lungs. III. The attack on immature lungs, *Am. Rev. Respir. Dis.,* 116, 497, 1977.
18. **Stent, G. S.,** Prematurity and uniqueness in scientific discovery, *Sci. Am.,* 227, 84, 1972.
19. **Comroe, J. H., Jr.,** in *Pulmonary and Respiratory Physiology,* Part 1, Comroe, J. H., Jr., Ed., Dowden, Hutchinson and Ross, Inc., Stroudsburg, Pa., 1976, 214.
20. **Macklin, C. C.,** Residual epithelial cells on the pulmonary walls of mammals, *Trans. R. Soc. Can. Sect.* 5, 40, 93, 1946.
21. **Staub, N. C., Clements, J. A., Permutt, S., and Proctor, D. F.,** Charles Clifford Macklin, 1883—1959: an appreciation, *Am. Rev. Respir. Dis.,* 114, 823, 1976.
22. **Macklin, C. C.,** The pulmonary alveolar mucoid film and the pneumonocytes, *Lancet,* i, 1099, 1954.
23. **Pattle, R. E.,** Properties, function and origin of the alveolar lining layer, *Nature (London),* 175, 1125, 1955.
24. **Pattle, R. E.,** Properties, function and origin of the alveolar lining layer, *Proc. R. Soc. London Ser. B,* 148, 217, 1958.
25. **Radford, E. P., Jr.,** Method for estimating respiratory surface area of mammalian lungs from their physical characteristics, *Proc. Soc. Exp. Biol. Med.,* 87, 58, 1954.
26. **Clements, J. A.,** Physiochemical properties of pulmonary surface films, *Proc. Int. Union Physiol. Sci.,* 1, 268, 1962.
27. **Clements, J. A.,** Dependence of pressure-volume characteristics of lungs on intrinsic surface active material, *Am. J. Physiol.,* 187, 592, 1956.
28. **Clements, J. A.,** Surface tension of lung extracts, *Proc. Soc. Exp. Biol. Med.,* 95, 170, 1957.
29. **Clements, J. A., Brown, E. S., and Johnson, R. P.,** Pulmonary surface tension and the mucus lining of the lungs: some theoretical considerations, *J. Appl. Physiol.,* 12, 262, 1958.
30. **Pattle, R. E. and Thomas, L. C.,** Lipoprotein composition of the film lining the lung, *Nature (London),* 189, 844, 1961.
31. **Buckingham, S.,** Studies on the identification of an antiatelectosis factor in normal sheep lung, *Am. J. Dis. Child.,* 102, 521, 1961.

32. **Klaus, M. H., Clements, J. A., and Havel, R. J.,** Composition of surface active material isolated from beef lung, *Proc. Natl. Acad. Sci. U.S.A.,* 47, 1858, 1961.
33. **Brown, E. S.,** Isolation and assay of dipalmitoyl lecithin in lung extracts, *Am. J. Physiol.,* 207, 402, 1964.
34. **Hochheim, K.,** Ueber einige Befunde in den Lungen von Neugeborenen und die Beziehung derselben zur Aspiration von Fruchtwasser, *Centralbl. Pathol.,* 14, 537, 1903.
35. **Johnson, W. C. and Meyer, J. R.,** A study of pneumonia in the stillborn and newborn, *Am. J. Obstet. Gynecol.,* 9, 151, 1925.
36. **Nelson, N. M.,** Historical perspective: past and present approaches to therapy in hyaline membrane disease, in *Lung Development and Clinical Perspectives,* Vol. 2, Farrell, P. M., Ed., Academic Press, New York, 1982, 3.
37. **Gruenwald, P.,** Surface tension as a factor in the resistance of neonatal lungs to aeration, *Am. J. Obstet. Gynecol.,* 53, 996, 1947.
38. **Avery, M. E. and Mead, J.,** Surface properties in relation to atelectosis and hyaline membrane disease, *Am. J. Dis. Child.,* 97, 517, 1959.
39. **Farrell, P. M. and Zachman, R. D.,** Pulmonary surfactant and the respiratory distress syndrome, in *Fetal and Maternal Medicine,* Quilligan, E. J. and Kretchmer, N., Eds., John Wiley & Sons, New York, 1980, 221.
40. **Schurch, S.,** Surface tension at low lung volumes: dependence on time and on alveolar size, *Respir. Physiol.,* 48, 339, 1982.
41. **McIver, D. J. L., Possmayer, F., and Schurch, S.,** A synthetic emulsion reproduces the functional properties of pulmonary surfactant, *Biochim. Biophys. Acta,* 751, 74, 1983.
42. **Kezdy, F. J.,** in *Membrane Molecular Biology,* Fox, C. F. and Keith, A. D., Eds., Sinauer Assoc. Inc., Stamford, Conn., 1972, 123.
43. **Sanders, R. L.,** The biochemical composition of the lung, in *Lung Development: Biological and Clinical Perspectives,* Vol. 1, Farrell, P. M., Ed., Academic Press, New York, 1982, 193.
44. **King, R. J.,** Isolation and chemical composition of pulmonary surfactant, in *Pulmonary Surfactant,* Robertson, B., van Golde, L. M. G., and Batenburg, J. J., Eds., Elsevier, Amsterdam, 1984, 1.
45. **Gil, J. and Reiss, O. K.,** Isolation and characterization of lamellar bodies and tubular myelin from rat lung homogenates, *J. Cell Biol.,* 58, 152, 1972.
46. **Magoon, M. W., Wright, J. R., Baritussio, A., Williams, M. C., Goerke, J., Benson, B. J., Hamilton, R. L., and Clements, J. A.,** Subfractionation of lung surfactant. Implications for metabolism and surface activity, *Biochim. Biophys. Acta,* 750, 18, 1983.
47. **Shelly, S. A., Balis, J. Y., Paciga, J. E., Espinoza, C. G., and Eichman, A. V.,** Biochemical composition of adult human lung surfactant, *Lung,* 160, 195, 1982.
48. **Harwood, J. L., Desai, R., Hext, P., Tetley, T., and Richards, R.,** Characterization of pulmonary surfactant from ox, rabbit, rat and sheep, *Biochem. J.,* 151, 707, 1975.
49. **Jobe, A.,** Surfactant phospholipid metabolism in 3-day and 3-day postmature rabbits in vivo, *Pediatr. Res.,* 14, 319, 1980.
50. **Yu, S., Harding, P. G. R., Smith, N., and Possmayer, F.,** Bovine pulmonary surfactant: chemical composition and physical properties, *Lipids,* 18, 522, 1983.
51. **Ohno, K., Akino, T., and Fujiwara, T.,** Phospholipid metabolism in perinatal lung, in *Reviews in Perinatal Medicine,* Vol. 2, Scarpelli, E. M. and Cosmi, E. V., Eds., Raven Press, New York, 1978, 227.
52. **Sato, T. and Akino, T.,** Source of lung surfactant phospholipids: comparison of palmitate and acetate as precursors, *Lipids,* 17, 884, 1982.
53. **Soodsma, J. F., Mims, L. C., and Harlow, R. D.,** The analysis of the molecular species of fetal rabbit lung phosphatidylcholine by consecutive chromatographic techniques, *Biochim. Biophys. Acta,* 424, 159, 1976.
54. **Post, M., Batenburg, J. J., Schuurmans, E. A. J. M., Laros, C. D., and van Golde, L. M. G.,** Lamellar bodies isolated from adult human lung tissue, *Exp. Lung Res.,* 3, 17, 1982.
55. **Sanders, R. L.,** Major phospholipids in pulmonary surfactant, in *Lung Development: Biological and Clinical Perspectives,* Vol. 1, Farrell, P. M., Ed., Academic Press, New York, 1982, 211.
56. **Abe, M. and Akino, T.,** Comparison of metabolic heterogeneity of glycerolipids in rat lung, *Tohoku J. Exp. Med.,* 106, 343, 1972.
57. **King, R. J.,** The apolipoproteins of pulmonary surfactant, *Prog. Respir. Res.,* 18, 68, 1984.
58. **King, R. J.,** Composition and metabolism of the apolipoproteins of pulmonary surfactant, *Ann. Rev. Physiol.,* 47, 775, 1985.
59. **Hawgood, S., Efrati, H., Schilling, J., and Benson, B. J.,** Chemical characterization of lung surfactant apoproteins: amino acid composition, N-terminal sequence and enzymatic digestion, *Biochem. Soc. Trans.,* 13, 1092, 1985.
60. **Lynn, W. S.,** Alveolyn, structure and source: a review, *Exp. Lung Res.,* 6, 191, 1984.
61. **Floros, J., Phelps, D. S., and Taeusch, H. W.,** Biosynthesis and *in vitro* translation of the major surfactant-associated protein from human lung, *J. Biol. Chem.,* 260, 495, 1985.

62. **Floros, J., Phelps, D. S., Kourembanas, S., and Taeusch, H. W.,** Primary translation products, biosynthesis, and tissue specificity of the major surfactant protein in rat, *J. Biol. Chem.,* 261, 828, 1986.
63. **Whitsett, J. A., Weaver, T., Hull, W., Ross, G., and Dion, C.,** Synthesis of surfactant-associated glycoprotein A by rat type II epithelial cells. Primary translation products and post-translational modification, *Biochim. Biophys. Acta,* 828, 162, 1985.
64. **Weaver, T. E., Hull, W. M., Ross, G. F., and Whitsett, J. A.,** Intracellular and oligomeric forms of surfactant-associated apolipoprotein(s) A in the rat, *Biochim. Biophys. Acta,* 827, 260, 1985.
65. **Weaver, T. E., Whitsett, J. A., Hull, W. M., and Ross, G.,** Identification of canine pulmonary surfactant-associated glycoprotein A precursors, *J. Appl. Physiol.,* 58, 2091, 1985.
66. **Whitsett, J. A., Ross, G., Weaver, T., Rice, W., Dion, C., and Hull, W.,** Glycosylation and secretion of surfactant-associated glycoprotein A, *J. Biol. Chem.,* 260, 15273, 1985.
67. **Ng, V. L., Herndon, V. L., Mendelson, C. R., and Snyder, J. M.,** Characterization of rabbit surfactant-associated proteins, *Biochim. Biophys. Acta,* 754, 218, 1983.
68. **Katyal, S. L. and Singh, G.,** Structural and ontogenetic relationships of rat lung surfactant apoproteins, *Exp. Lung Res.,* 6, 175, 1984.
69. **Benson, B., Hawgood, S., Schilling, J., Clements, J., Damm, D., Cordell, B., and Tyler, R. T.,** Structure of canine pulmonary surfactant apoprotein: cDNA and complete amino acid sequence, *Proc. Natl. Acad. Sci. U.S.A.,* 82, 6379, 1985.
70. **White, R. T., Damm, D., Miller, J., Spratt, K., Schilling, J., Hawgood, S., Benson, B., and Cordell, B.,** Isolation and characterization of the human pulmonary surfactant apoprotein gene, *Nature (London),* 317, 361, 1985.
71. **Thurlbeck, W. M.,** Postnatal growth and development of the lung, *Am. Rev. Respir. Dis.,* 111, 803, 1975.
72. **Kikkawa, Y.,** Morphology and morphologic development of the lung, in *Pulmonary Physiology of the Fetus, Newborn and Child,* Scarpelli, E. M., Ed., Lea & Febiger, Philadelphia, 1975, 37.
73. **Williams, M.,** Development of the alveolar structure of the fetal rat in late gestation, *Fed. Proc.,* 36, 2653, 1977.
74. **Stahlman, M. T. and Gray, M. E.,** Anatomical development and maturation of the lungs, *Clin. Perinatol.,* 5, 181, 1978.
75. **Farrell, P. M.,** Morphologic aspects of lung maturation, in *Lung Development: Biological and Clinical Perspectives,* Vol. 1, Farrell, P. M., Ed., Academic Press, New York, 1982, 13.
76. **Kuhn, C., III,** The cytology of the lung: ultrastructure of the respiratory epithelium and extracellular lining layers, in *Lung Development: Biological and Clinical Perspectives,* Vol. 1, Farrell, P. M., Ed., Academic Press, New York, 1982, 27.
77. **Stratton, C. J.,** Morphology of surfactant producing cells and of the alveolar lining layer, in *Pulmonary Surfactant,* Robertson, B., van Golde, L. M. G., and Batenburg, J. J., Eds., Elsevier, Amsterdam, 1984, 67.
78. **Gil, J.,** Histological preservation and ultrastructure of alveolar surfactant, *Ann. Rev. Physiol.,* 47, 753, 1985.
79. **Hook, G. E. R.,** Extracellular hydrolases of the lung, *Biochemistry,* 17, 520, 1978.
80. **Hook, G. and Gilmore, L. B.,** Hydrolases of pulmonary lysosomes and lamellar bodies, *J. Biol. Chem.,* 257, 9211, 1982.
81. **de Vries, A. C. J., Schram, A. W., Tager, J. M., Batenburg, J. J., and van Golde, L. M. G.,** A specific alpha-glucosidase in lamellar bodies of the human lung, *Biochim. Biophys. Acta,* 837, 230, 1985.
82. **Goerke, J.,** Lung surfactant, *Biochim. Biophys. Acta,* 344, 241, 1974.
83. **Benson, B. J., Hawgood, S., and Williams, M. C.,** Role of surfactant proteins in surfactant structure and function, *Exp. Lung Res.,* 6, 223, 1984.
84. **Wright, J. R., Benson, B. J., Williams, M. C., Goerke, J., and Clements, J. A.,** Protein composition of rabbit alveolar surfactant subfractions, *Biochim. Biophys. Acta,* 791, 320, 1984.
85. **Mason, R. J.,** Pulmonary alveolar type II epithelial cells and adult respiratory distress syndrome, *West. J. Med.,* 143, 611, 1985.
86. **Haagsman, H. P. and van Golde, L. M. G.,** Lung surfactant and pulmonary toxicology, *Lung,* 163, 275, 1985.
87. **Clements, J. A. and Tooley, W. H.,** kinetics of surface-active material in the fetal lung, in *Development of the Lung,* Hodson, W. A., Ed., Marcel Dekker, New York, 1977, 349.
88. **Jost, A. and Policard, A.,** Contribution experimentale a l'etude du developpement prenatal du poumon chez le lapin, *Arch. Anat. Microsc.,* 37, 323, 1948.
89. **Hallman, M.,** Antenatal diagnosis of lung maturity, in *Pulmonary Surfactant,* Robertson, B., van Golde, L. M. G., and Batenburg, J. J., Eds., Elsevier, Amsterdam, 1984, 419.
90. **Tsao, F. H. C. and Zachman, R. D.,** Prenatal assessment of fetal lung maturation: a critical review of amniotic fluid phospholipid tests, in *Lung Development: Biological and Clinical Perspectives,* Vol. 2, Farrell, P. M., Ed., Academic Press, New York, 1982, 167.

91. **Gluck, L., Kulovich, M. V., Borer, R. C., Brenner, P. H., Anderson, G. G., and Spellacy, W. N.,** Diagnosis of the respiratory distress syndrome by amniocentesis, *Am. J. Obstet. Gynecol.,* 109, 440, 1971.

92. **Spillman, T., Cotton, D. B., Lynn, S. C., Jr., and Bretandiere, J.-P.,** Influence of phospholipid saturation on classical thin-layer chromatographic methods and its effect on amniotic fluid lecithin/sphingomyelin ratio determinations, *Clin. Chem.,* 29, 250, 1983.

93. **Mason, R., Nellenbogen, J., and Clements, J. A.,** Isolation of disaturated phosphatidylcholine with osmium tetroxide, *J. Lipid Res.,* 17, 281, 1976.

94. **Hallman, M., Feldman, B. H., Kirkpatrick, E., and Gluck, L.,** Absence of phosphatidylglycerol (PG) in respiratory distress syndrome in the newborn. Study of the minor surfactant phospholipids in newborns, *Pediatr. Res.,* 11, 714, 1977.

95. **Kulovich, M. V., Hallman, M., and Gluck, L.,** The lung profile. I. Normal pregnancy, *Am. J. Obstet. Gynecol.,* 135, 57, 1979.

96. **Kulovich, M. V. and Gluck, L.,** The lung profile. II. Complicated pregnancy, *Am. J. Obstet. Gynecol.,* 135, 64, 1979.

97. **King, R. J., Ruch, J., Gikas, E., Platzker, A. C. G., and Creasy, R. K.,** Appearance of apoproteins of pulmonary surfactant in human amniotic fluid, *J. Appl. Physiol.,* 39, 735, 1975.

98. **Shelley, S. A., Balis, J. U., Paciga, J. E., Knuppel, R. A., Ruffolo, E. H., and Bouis, P. J.,** Surfactant "apoproteins" in human amniotic fluid: an enzyme-linked immunosorbent assay for the prenatal assessment of lung maturity, *Am. J. Obstet. Gynecol.,* 144, 224, 1982.

99. **Katyal, S. L. and Singh, G.,** An enzyme-linked immunoassay of surfactant proteins. Its application to the study of fetal lung development in the rat, *Pediatr. Res.,* 17, 439, 1983.

100. **Kuroki, Y., Takahashi, H., Fukada, Y., Mikawa, M., Inagawa, A., Fujimoto, S., and Akino, T.,** Two-site "simultaneous" immunoassay with monoclonal antibodies for the detection of surfactant apoproteins in human amniotic fluid, *Pediatr. Res.,* 19, 1017, 1985.

101. **Schleuter, M., Phibbs, R. H., Creasy, R. K., Clements, J. A., and Tooley, W. H.,** Antenatal prediction of graduated risk of hyaline membrane disease by amniotic fluid foam test for pulmonary surfactant, *Am. J. Obstet. Gynecol.,* 134, 761, 1979.

102. **Shinitzky, M., Goldfisher, A., Bruck, A., Goldman, B., Stern, E., Barkai, G., Mashiach, S., and Serr, D. M.,** A new method for assessment of fetal lung maturity, *Br. J. Obstet. Gynecol.,* 83, 838, 1976.

103. **Stark, R. I., Blumenfeld, T. A., Cheskin, H. S., Dryenfurth, I., and James, L. S.,** Amniotic fluid fluorescence polarization value as a predictor of respiratory distress syndrome, *J. Pediatr.,* 96, 301, 1980.

104. **O'Brien, W. F. and Cefalo, R. C.,** Clinical applicability of amniotic fluid tests for fetal pulmonary fluid, *Am. J. Obstet. Gynecol.,* 136, 135, 1980.

105. **Butterworth, J., Broadhead, D. M., Sutherland, G. R., and Bain, G. R.,** Liposomal enzymes of amniotic fluid in relation to gestational age, *Am. J. Obstet. Gynecol.,* 119, 821, 1974.

106. **Bleasdale, J. E., Davis, C., and Agranoff, B. W.,** The measurement of phosphatidate phosphohydrolase in human amniotic fluid, *Biochim. Biophys. Acta,* 528, 331, 1978.

107. **Possmayer, F.,** The perinatal lung, in *The Fetus and the Neonate,* Jones, C. T., Ed., Elsevier/North Holland, Amsterdam, 1982, 337.

108. **Ballard, P. L.,** Hormonal aspects of fetal lung development, in *Lung Development: Biological and Clinical Perspectives,* Vol. 2, Farrell, P. M., Ed., Academic Press, New York, 1982, 205.

109. **Smith, B. T.,** Pulmonary surfactant during fetal development and neonatal adoption: hormonal control, in *Pulmonary Surfactant,* Robertson, B., van Golde, L. M. G., and Batenburg, J. J., Eds., Elsevier, Amsterdam, 1984, 357.

110. **Avery, M. E.,** Prevention of neonatal RDS by pharmacological methods, in *Pulmonary Surfactant,* Robertson, R., van Golde, L. M. G., and Batenburg, J. J., Eds., Elsevier, Amsterdam, 449, 1982.

111. **Rooney, S. A.,** The surfactant system and lung phospholipid biochemistry, *Am. Rev. Resp. Dis.,* 131, 439, 1985.

112. **Perelman, R. H., Farrell, P. M., Engle, M. J., and Kemnitz, J. W.,** Developmental aspects of lung lipids, *Annu. Rev. Physiol.,* 47, 803, 1985.

113. **Liggins, G. C.,** Premature delivery of fetal lambs infused with glucocorticoids, *J. Endocrinol.,* 45, 515, 1969.

114. **Smith, B. T., Torday, J. S., and Giroud, C. J. P.,** Evidence for different gestation-dependent effects of cortisol on cultured fetal lung cells, *J. Clin. Invest.,* 53, 1518, 1974.

115. **Gross, I., Dynia, D. W., Wilson, C. M., Ingleson, L. D., Gewolb, I. H., and Rooney, S. A.,** Glucocorticoid-thyroid hormone interactions in fetal rat lung, *Pediatr. Res.,* 18, 191, 1984.

116. **Snyder, J. M., Johnston, J. M., and Mendelson, C. R.,** Differentiation of type II cells of human lung *in vitro, Cell. Tissue Res.,* 220, 17, 1981.

117. **Gross, I., Ballard, P. L., Ballard, R. A., Jones, C. T., and Wilson, C. M.,** Corticosteroid stimulation of phosphatidylcholine synthesis in cultured fetal rabbit lung: evidence for *de novo* protein synthesis mediated by glucocorticoid receptors, *Endocrinology,* 112, 829, 1983.

118. **Martin, C. E., Cake, M. H., Hartman, P. E., and Cook, I. F.,** Relationship between foetal cortico-steroids, maternal progesterone and parturition in the rat, *Acta Endocrinol.,* 84, 167, 1977.
119. **Barr, H. A., Lugg, M. A., and Nicholas, T. E.,** Cortisone and cortisol in maternal and fetal blood and in amniotic fluid during the final ten days of gestation in the rabbit, *Biol. Neonate,* 38, 214, 1980.
120. **Murphy, B. E. P.,** Cortisol and cortisone levels in the cord blood at delivery of infants with and without the respiratory distress syndrome, *Am. J. Obstet. Gynecol.,* 119, 1112, 1974.
121. **Smith, I. D. and Shearman, R. P.,** The relationship of human umbilical arterial and venous plasma levels of corticosteroids to gestational age, *J. Obstet. Gynecol. Br. Common.,* 81, 11, 1974.
122. **Blackburn, W. R., Kelly, J. S., Dickman, P. S., Travers, H., Lopata, M. A., and Rhoades, R. A.,** The role of the pituitary-adrenal-thyroid axis in lung differentiation. II. Biochemical studies of developing lung in anencephalic fetal rats, *Lab. Invest.,* 28, 352, 1973.
123. **Meyrick, B., Bearn, J. G., Cogg, A. G., Monkhouse, C. R., and Reid, L.,** The effect of in utero decapitation on the morphological and physiological development of the rabbit lung, *J. Anat.,* 119, 517, 1975.
124. **Farrell, P. M. and Avery, M. E.,** Hyaline membrane disease, *Am. Rev. Resp. Dis.,* 111, 657, 1975.
125. **Shields, J. R. and Resnik, R.,** Fetal lung maturation and the antenatal use of glucocorticoids to prevent the respiratory distress syndrome, *Obstet. Gynaecol. Rev.,* 34, 343, 1979.
126. **Liggins, G. C. and Howie, R. N.,** A controlled trial of antepartum glucocorticoid treatment for prevention of the respiratory distress syndrome in premature infants, *Pediatrics,* 50, 515, 1972.
127. Collaborative group on antenatal steroid therapy, Effect of antenatal dexamethasone administration on the prevention of respiratory distress syndrome, *Am. J. Obstet. Gynecol.,* 141, 276, 1981.
128. **Taeusch, H. W.,** Glucocorticoid prophylaxis for respiratory distress syndrome: a review of potential toxicity, *J. Pediatr.,* 87, 617, 1975.
129. **Sanfacon, R., Possmayer, F., and Harding, P. G. R.,** Dexamethasone treatment of the guinea pig fetus: its effects on the incorporation of ^{3}H thymidine into deoxyribonucleic acid, *Am. J. Obstet. Gynecol.,* 127, 745, 1977.
130. **Quirk, J. G., Raker, R. K., Petrie, R. H., and Williams, A. M.,** The role of glucocorticoids, unstressful labor, and atraumatic delivery in the prevention of respiratory distress syndrome, *Am. J. Obstet. Gynecol.,* 134, 768, 1979.
131. **Kauffman, S. L.,** Acceleration of canalicular development in lungs of fetal mice exposed transplacentally to dexamethasone, *Lab. Invest.,* 36, 395, 1977.
132. **Beck, J. C., Mitzner, W., Johnson, J. W. C., Hutchins, G. M., Foidart, J.-M., London, W. T., Palmer, A. E., and Scott, R.,** Betamethasone and the rhesus fetus: effect on lung morphology and connective tissue, *Pediatr. Res.,* 15, 235, 1981.
133. **Patrick, J., Challis, J., Campbell, K., Carmichael, L., Richardson, B., and Tevaarwerk, G.,** Effects of synthetic glucocorticoid administration on human fetal breathing movements at 34 to 35 weeks gestation, *Am. J. Obstet. Gynecol.,* 139, 324, 1981.
134. **Alcorn, D., Adamson, T. M., Maloney, J. E., and Robinson, P. M.,** Morphological effects of chronic bilateral phrenectomy or vagotomy in the fetal lamb, *J. Anat.,* 130, 683, 1980.
135. **Post, M., Floros, J., and Smith, B. T.,** Inhibition of lung maturation by monoclonal antibodies raised against fibroblast pneumonocyte factor, *Nature (London),* 308, 284, 1984.
136. **Wu, B., Kikkawa, Y., Orzalesi, M. M., Motoyama, E. K., Kaibara, M., Zigas, C. J., and Cook, C. D.,** The effect of thyroxin on the maturation of fetal rabbit lungs, *Biol. Neonate,* 22, 161, 1973.
137. **Ballard, P. L., Benson, B. J., Brehier, A,. Carter, J. P., Kriz, B. M., and Jorgensen, E. C.,** Transplacental stimulation of lung development in the fetal rabbit by 3,5-dimethyl-3'-isopropyl-L-thyronine, *J. Clin. Invest.,* 65, 1407, 1980.
138. **Cunningham, M. D., Hollingsworth, D. R., and Belin, R. P.,** Impaired surfactant production in Cretin lambs, *Obstet. Gynecol.,* 55, 439, 1980.
139. **Morishige, W. K.,** Thyroid hormones influence glucocorticoid receptor levels in the neonatal rat lung, *Endocrinology,* 111, 1017, 1982.
140. **Smith, B. T. and Sabry, K.,** Glucocorticoid-thyroid synergism in lung maturation: a mechanism involving epithelial-mesenchymal interaction, *Proc. Natl. Acad. Sci. U.S.A.,* 80, 1951, 1983.
141. **Gonzales, L. W. and Ballard, P. L.,** Nuclear 3,5,3'-triiodothyronine receptors in rabbit lung: characterization and developmental changes, *Endocrinology,* 111, 542, 1982.
142. **Abdul-Karim, R. W., Prior, J. T., and Havilland, M. E.,** The influence of estrogens on the lung vasculature of the premature rabbit neonate, *J. Reprod.,* 11, 140, 1969.
143. **Mendelson, C. R., MacDonald, P. C., and Johnston, J. M.,** Estrogen binding in human fetal lung tissue cytosol, *Endocrinology,* 106, 368, 1980.
144. **Mendelson, C. R., Brown, P. K., MacDonald, P. C., and Johnston, J. M.,** Characterization of a cytosolic estrogen-binding protein in lung tissue of fetal rats, *Endocrinology,* 109, 210, 1981.
145. **van Petten, G. R. and Bridges, R.,** The effects of prolactin on pulmonary maturation in the fetal rabbit, *Am. J. Obstet. Gynecol.,* 134, 711, 1979.

146. **Cox, M. A. and Torday, J. S.,** Pituitary oligopeptide regulation of phosphatidylcholine synthesis by fetal rabbit lung cells: lack of effect with prolactin, *Am. Rev. Respir. Dis.,* 123, 181, 1981.
147. **Ballard, P. L., Gluckman, P. D., Brehier, A., Kitterman, J. A., Kaplan, S. L., Rudolph, A. M., and Grumbach, M. M.,** Failure to detect an effect of prolactin on pulmonary surfactant and adrenal steroids in fetal sheep and rabbits, *J. Clin. Invest.,* 62, 879, 1978.
148. **Mendelson, C. R., Johnston, J. M., MacDonald, P. C., and Snyder, J. M.,** Multihormonal regulation of surfactant synthesis by human fetal lung *in vitro, J. Clin. Endocrinol. Metab.,* 53, 307, 1981.
149. **Snyder, J. M., Longmuir, K. J., Johnston, J. M., and Mendelson, C. R.,** Hormonal regulation of the synthesis of lamellar body phosphatidylglycerol and phosphatidylinositol in fetal lung tissue, *Endocrinology,* 112, 1012, 1983.
150. **Torday, J. S., Nielsen, H. C., Fencl, M., de M., and Avery, M. E.,** Sex differences in fetal lung maturation, *Am. Rev. Respir. Dis.,* 123, 205, 1981.
151. **Nielsen, H. C., Zinman, H. M., and Torday, J. S.,** Dihydrotestosterone inhibits fetal rabbit pulmonary surfactant production, *J. Clin. Invest.,* 69, 611, 1982.
152. **Hallman, M.,** Induction of surfactant phosphatidylglycerol in the lung of fetal and newborn rabbits by dibutyryl adenosine 3',5' monophosphate, *Biochem. Biophys. Res. Commun.,* 77, 1094, 1977.
153. **Whitsett, J. A., Noguchi, A., and Moore, J. J.,** Developmental aspects of alpha- and beta-adrenergic receptors, *Semin. Perinat.,* 6, 125, 1982.
154. **Robert, M. F., Neff, R. K., Hubbell, J. P., Taeusch, H. W., and Avery, M. E.,** Association between maternal diabetes and the respiratory distress syndrome in the newborn, *New Engl. J. Med.,* 294, 357, 1976.
155. **Bose, C. L., Manne, D. N., D'Ercole, A. J., and Lawson, E. E.,** Delayed fetal pulmonary maturation in a rabbit model of the diabetic pregnancy, *J. Clin. Invest.,* 66, 220, 1980.
156. **Sosenko, I. R. S., Lawson, E. E., Demottaz, V., and Frantz, I. D.,** Functional delay of lung maturation in fetuses of diabetic rabbits, *J. Appl. Physiol.,* 48, 643, 1980.
157. **Neufeld, N. D., Sevanian, A., Barrett, C. T., and Kaplan, S. A.,** Inhibition of surfactant production by insulin in fetal rabbit lung slices, *Pediatr. Res.,* 13, 752, 1979.
158. **Gross, I., Smith, G. J. W., Wilson, C. M., Maniscalco, W. M., Ingleson, L. D., Brehier, A., and Rooney, S. A.,** The influence of hormones in the biochemical development of fetal rat lung in organ culture. II. Insulin, *Pediatr. Res.,* 14, 834, 1980.
159. **Sosenko, I. R. S., Hartig-Beeken, I., and Frantz, I. D., III,** Cortisol reversal of functional delay of lung maturation in fetuses of diabetic rabbits, *J. Appl. Physiol.,* 49, 971, 1980.
160. **Mulay, S. and Solomon, S.,** Influence of streptozotocin-induced diabetes in pregnant rats on plasma corticosterone and progesterone levels and on cytoplasmic glucocorticoid receptors in fetal tissues, *J. Endocrinol.,* 96, 335, 1983.
161. **Rooney, S. A.,** Biochemical development of the lung, in *The Biological Basis of Reproduction and Developmental Medicine,* Warshaw, J. B., Ed., Elsevier, Amsterdam, 1983, 239.
162. **Hollingsworth, M. and Gilfillan, A. M.,** The pharmacology of lung surfactant secretion, *Pharmacol. Rev.,* 36, 69, 1984.
163. **Nylund, L., Lagercrantz, H., and Lanell, N. O.,** Catecholamines in fetal blood during birth in man, *J. Dev. Physiol.,* 1, 427, 1979.
164. **Lagercrantz, H. and Bistoletti, P.,** Catecholamine release in the newborn infant at birth, *Pediatr. Res.,* 11, 889, 1977.
165. **Bistoletti, P., Nylund, L., Lagercrantz, H., Hjemdahl, P., and Strom, H.,** Fetal scalp catecholamines during labour, *Am. J. Obstet. Gynecol.,* 147, 785, 1983.
166. **Lagercrantz, H. and Slotkin, T. A.,** The "stress" of being born, *Sci. Am.,* 254, 100, 1986.
167. **Lawson, E. E., Brown, E. R., Torday, M. S., Madansky, D. L., and Taeusch, H. W., Jr.,** The effect of epinephrine on tracheal fluid flow and surfactant efflux in fetal sheep, *Am. J. Respir. Dis.,* 118, 1023, 1978.
168. **Robertson, B.,** Neonatal mechanics and morphology after experimental therapeutic regimens, in *Reviews in Perinatal Medicine,* Vol. 4, Scarpelli, E. M. and Cosmi, E. V., Eds., Raven Press, New York, 1981, 337.
169. **Fisher, A. B. and Chander, A.,** Intracellular processing of surfactant lipids in the lung, *Ann. Rev. Physiol.,* 47, 789, 1985.
170. **Whitsett, J. A., Hull, W., Dion, C., and Lessard, J.,** c-AMP dependent actin phosphorylation in developing rat lung and type II epithelial cells, *Exp. Lung Res.,* 9, 191, 1985.
171. **Whitsett, J. A., Matz, S., and Darovec-Beckerman, C.,** c-AMP dependent protein kinase and protein phosphorylation in developing rat lung, *Pediatr. Res.,* 17, 959, 1983.
172. **Giannopoulos, G.,** Identity and ontogeny of beta-adrenergic receptors in fetal rabbit lung, *Biochem. Biophys. Res. Commun.,* 95, 388, 1980.

173. **Roberts, J. M. and Musci, T. J.,** Alveolar beta-adrenoceptors: modulation and role in perinatal adaptation, in *Respiratory Control and Lung Development in the Fetus and Newborn,* Gluckman, P. and Johnson, B., Eds., Perinatology Press, Ithaca, N.Y., 1986, 135.

174. **Whitsett, J. A., Machulskis, A., Noguchi, A., and Bardsall, J. A.,** Ontogeny of alpha- and beta-adrenergic receptors in rat lung, *Life Sci.,* 30, 139, 1982.

175. **Giannopoulos, G. and Smith, S. K.,** Hormonal regulation of beta-adrenergic receptors in fetal rabbit lung in organ culture, *Life Sci.,* 31, 795, 1982.

176. **Maniscalco, W. M. and Shapiro, D. L.,** Effects of dexamethasone on beta-adrenergic receptors in fetal lung explants, *Pediatr. Res.,* 17, 274, 1983.

177. **Barnes, P., Jacobs, M., and Roberts, J. M.,** Glucocorticoids preferentially increase fetal alveolar beta-adrenoreceptors, *Pediatr. Res.,* 18, 1191, 1984.

178. **Moawael, A.-H., River, P., and Lin, C.,** Estrogen increases beta-adrenergic binding in the preterm fetal rabbit lung, *Am. J. Obstet. Gynecol.,* 151, 514, 1985.

179. **Das, D. K., Ayromlooi, J., Bandyopayhyay, S., Bandyopadhya, A., Neogi, A., and Steinberg, H.,** Potentiation of surfactant release in fetal lung by thyroid hormone action, *J. Appl. Physiol.,* 56, 1621, 1984.

180. **Padbury, J. F., Jacobs, H. C., Lam, R. W., Conaway, D., Jobe, A. H., and Fisher, D. A.,** Adrenal epinephrine and the regulation of pulmonary surfactant release in neonatal rabbits, *Exp. Lung Res.,* 7, 177, 1984.

181. **Robertson, B.,** Surfactant substitution: experimental models and clinical applications, *Lung,* 158, 57, 1980.

182. **Enhorning, G., Chamberlain, D., Contreras, C., Burgoyne, R., and Robertson, B.,** Isoxsuprine infusion to the pregnant rabbit and its effect on fetal lung surfactant, *Biol. Neonate,* 35, 43, 1979.

183. **Brown, M. J., Olver, R. E., Ramsden, C. A., Strang, L. B., and Walters, D. V.,** Effects of adrenaline and of spontaneous labour on the secretion and adsorption of lung liquid in the fetal lamb, *J. Physiol.,* 344, 132, 1983.

184. **Brown, M. J., Olver, R. E., Ramsden, C. A., Strang, L. B., and Walters, D. V.,** Pulmonary circulation and lung fluid balance in lambs during mechanical ventilation at different frequencies and inflation volumes, *J. Physiol.,* 344, 137, 1983.

185. **Olver, R. E.,** Fetal lung liquids, *Fed. Proc.,* 36, 2669, 1977.

186. **Bland, R. D., Hansen, T. N., Itaberkern, C. M., Bressack, M. A., Hazinski, T. A., Raj, J. U., and Goldberg, R. B.,** Lung fluid balance in lambs before and after birth, *J. Appl. Physiol.,* 53, 992, 1982.

187. **Notter, R. H. and Finkelstein, J. N.,** Pulmonary surfactant: an interdisciplinary approach, *J. Appl. Physiol. Respir. Environ. Exercise Physiol.,* 57, 1613, 1984.

188. **Notter, R. H.,** Surface chemistry of pulmonary surfactant: the role of individual components, in *Pulmonary Surfactant,* Robertson, B., van Golde, L. M. G., and Batenburg, J. J., Eds., Elsevier, Amsterdam, 1984, 17.

189. **Possmayer, F., Yu, S.-H., Weber, M., and Harding, P. G. R.,** Pulmonary surfactant, *Can. J. Biochem. Cell Biol.,* 62, 1121, 1984.

190. **King, R. J. and Clements, J. A.,** Surface active materials from dog lungs. II. Composition and physiological correlations, *Am. J. Physiol.,* 223, 715, 1972.

191. **Benson, B. J., Williams, M. C., Hawgood, S., and Sargeant, T.,** Role of lung surfactant specific proteins in surfactant structure and function, *Prog. Respir. Res.,* 18, 83, 1984.

192. **Benson, B. J., Williams, M. C., Sueishi, K., Goerke, J., and Sargent, T.,** Role of calcium ions in the structure and function of pulmonary surfactant, *Biochim. Biophys. Acta,* 793, 18, 1983.

193. **Hawgood, S., Benson, B. J., and Hamilton, R. L., Jr.,** Effects of a surfactant-associated protein and calcium ions on the structure and surface activity of lung surfactant lipids, *Biochemistry,* 24, 184, 1984.

194. **Enhorning, G.,** Pulsating bubble technique for evaluating pulmonary surfactant, *J. Appl. Physiol.,* 43, 198, 1977.

195. **Metcalfe, I. L., Enhorning, G., and Possmayer, F.,** Pulmonary surfactant associated proteins: their role in the expression of surface activity, *J. Appl. Physiol.,* 49, 34, 1980.

196. **Suzuki, Y., Nakai, E., and Ohkawa, K.,** Experimental studies on the pulmonary surfactant reconstitution of surface-active material, *J. Lipid Res.,* 23, 53, 1982.

197. **Fujiwara, T.,** Surfactant replacement in neonatal RDS, in *Pulmonary Surfactant,* Robertson, B., van Golde, L. M. G., and Batenburg, J. J., Eds., Elsevier, Amsterdam, 1984, 479.

198. **Yu, S.-H. and Possmayer, F.,** Reconstitution of surfactant activity using the 5-6 K apoprotein associated with pulmonary surfactant, *Biochem. J.,* 236, 85, 1986.

199. **Whitsett, J. A., Notter, R. H., Ohnung, B. L., Ross, G., Meuth, J., Holm, B. A., Shapiro, D. L., and Weaver, T. E.,** Hydrophobic 6,000 kilodalton protein and its importance for biophysical activity in lung surfactant extracts, *Pediatr. Res.,* 20, 460, 1986.

200. **Taeusch, H. W., Keough, K. M. W., Williams, M., Slavin, R., Steele, G., Lee, A. S., Phelps, D., Kariel, V., Floros, J., and Avery, M. E.,** Characterization of bovine surfactant for infants with Respiratory Distress Syndrome, *Pediatrics ,* 77, 572, 1986.

201. **Whitsett, J. A., Hull, W. H., Ohning, B., Ross, G., and Weaver, T. E.,** Immunological identification of a pulmonary surfactant-associated protein of M_r = 6200 daltons, *Pediatr. Res.,* 20, 744, 1986.
202. **Metcalfe, I. L., Burgoyne, R., and Enhorning, G.,** Surfactant supplementation in the preterm rabbit: effects of applied volume on compliance and survival, *Pediatr. Res.,* 16, 834, 1982.
203. **Metcalfe, I. L., Pototschnik, R., Burgoyne, R., and Enhorning, G.,** Lung expansion and survival in rabbit neonates treated with surfactant extracts, *J. Appl. Physiol.,* 53, 838, 1982.
204. **Berggren, P., Curstedt, T., Grossman, G., Nilsson, R., and Robertson, B.,** Physiological activity of pulmonary surfactant with low protein content: effect of enrichment with synthetic phospholipids, *Exp. Lung Res.,* 8, 29, 1985.
205. **Claypool, W. D., Wang, D. L., Chander, A., and Fisher, A. B.,** An ethanol/ether soluble apoprotein from rat lung surfactant augments liposome uptake by isolated granular pneumocytes, *J. Clin. Invest.,* 74, 677, 1984.
206. **Robertson, B.,** Lung surfactant for replacement therapy, *Clin. Physiol.,* 3, 97, 1983.
207. **Robertson, B.,** Pathology and pathophysiology of neonatal surfactant deficiency, in *Pulmonary Surfactant,* Robertson, B., van Golde, L. M. G., and Batenburg, J. J., Eds., Elsevier, Amsterdam, 1984, 383.
208. **Enhorning, G. and Robertson, B.,** Lung expansion in the premature rabbit fetus after tracheal deposition of surfactant, *Pediatrics,* 50, 58, 1972.
209. **Enhorning, G., Grossmann, G., and Robertson, B.,** Tracheal deposition of surfactant before the first breath, *Am. Rev. Resp. Dis.,* 107, 921, 1973.
210. **Enhorning, G., Grossmann, G., and Robertson, B.,** Pharyngeal deposition of surfactant in the premature rabbit fetus, *Biol. Neonate,* 22, 126, 1973.
211. **Enhorning, G., Robertson, B., Milne, E., and Wagner, R.,** Radiologic evaluation of the premature newborn rabbit after pharyngeal deposition of surfactant, *Am. J. Obstet. Gynecol.,* 121, 475, 1975.
212. **Enhorning, G.,** Photography of peripheral pulmonary airway expansion as affected by surfactant, *J. Appl. Physiol.,* 42, 976, 1977.
213. **Grossman, G.,** Expansion pattern of terminal air-spaces in the premature rabbit lung after tracheal deposition of surfactant, *Pflugers Arch.,* 367, 205, 1977.
214. **Robertson, B. and Enhorning, G.,** The alveolar lining layer of the premature newborn rabbit after pharyngeal deposition of surfacant, *Lab. Invest.,* 31, 54, 1974.
215. **Enhorning, G., Hill, D., Sherwood, G., Cutz, E., Robertson, B., and Bryan, C.,** Improved ventilation of prematurely-delivered primates following tracheal deposition of surfactant, *Am. J. Obstet. Gynecol.,* 132, 529, 1978.
216. **Nilsson, R., Grossmann, G., and Robertson, B.,** Lung surfactant and the pathogenesis of neonatal bronchiolar lesions induced by artificial ventilation, *Pediatr. Res.,* 12, 249, 1978.
217. **Cutz, E., Enhorning, G., Robertson, B., Sherwood, W. G., and Hill, D. E.,** Effect of the surfactant prophylaxis on lung morphology in premature primates, *Am. J. Pathol.,* 92, 581, 1978.
218. **Adams, F. H., Towers, B., Osher, A. B., Ikegami, M., Fujiwara, T., and Nozaki, M.,** Effect of tracheal instillation of natural surfactant in premature lambs. I. Clinical and autopsy findings, *Pediatr. Res.,* 12, 841, 1978.
219. **Jobe, A., Ikegami, M., Glatz, T., Yoshida, Y., Diakomanolis, E., and Padbury, J.,** Saturated phosphatidylcholine secretion and the effect of natural surfactant on premature and term lambs ventilated for two days, *Exp. Lung Res.,* 4, 259, 1983.
220. **Egan, E. A., Notter, R. H., Shapiro, D. L., and Kwong, M. S.,** Natural and artificial lung surfactant replacement therapy in premature lambs, *J. Appl. Physiol.,* 55, 875, 1983.
221. **Fujiwara, T., Maeta, H., Chida, S., Morita, T., Watabe, Y., and Abe, T.,** Artificial surfactant therapy in hyaline membrane disease, *Lancet,* i, 55, 1980.
222. **Fujiwara, T. and Adams, F. H.,** Surfactant for hyaline membrane disease, *Pediatrics,* 66, 795, 1980.
223. **Smyth, J. A., Metcalfe, I. L., Duffty, P., Possmayer, F., Bryan, M. H., and Enhorning, G.,** Treatment of hyaline membrane disease with bovine surfactant, *Pediatrics,* 71, 913, 1983.
224. **Hallman, M., Merritt, A., Schneider, H., Epstein, B. L., Mannino, F., Edwards, D. K., and Gluck, L.,** Isolation of human surfactant from amniotic fluid and a pilot study of its efficacy in Respiratory Distress Syndrome, *Pediatrics,* 71, 473, 1983.
225. **Robertson, G., Berggren, P., Curstedt, T., Grossman, G., Herin, P., Ingelman-Sundberg, H., Mortensson, W., Nilsson, R., Noack, G., and Nohara, K.,** Clinical trials of natural surfactant phospholipids in the treatment of the Neonatal Respiratory Distress Syndrome (RDS), in *Ross Laboratories Special Conference: Clinical Trials on the Use of Surfactant in the Neonatal Respiratory Distress Syndrome,* Lacey, J. F., Ed., Ross Laboratories, Columbus, Ohio, 1984, 145.
226. **Lachman, B., Berggren, P., Curstedt, T., Grossman, G., and Robertson, B.,** Combined effects of surfactant substitution and prolongation of inspiration phase in artificially ventilated premature newborn rabbits, *Pediatr. Res.,* 16, 921, 1982.
227. **Ikegami, M., Jobe, A., Jacobs, H., and Lam, R.,** A protein from airways of premature lambs that inhibits surfactant function, *J. Appl. Physiol.,* 57, 1134, 1984.

228. **Morley, C. J., Bangham, A. D., Johnson, P., Thorburn, G. D., and Jenkin, G.,** Physical and physiological properties of dry lung surfactant, *Nature (London),* 271, 162, 1978.

229. **Bangham, A. D., Morley, C. J., and Phillips, M. C.,** The physical properties of an effective lung surfactant. *Biochim. Biophys. Acta,* 573, 552, 1979.

230. **Morley, C., Robertson, B., Lachmann, B., Nilsson, R., Bangham, A., Grossman, G., and Miller, N.,** Artificial surfactant and natural surfactant: comparative study of the effects on premature rabbit lungs, *Arch. Dis. Child.,* 55, 758, 1980.

231. **Morley, C. J., Bangham, A. D., Miller, N., and Davis, J. A.,** Dry artificial lung surfactant and its effect on very small premature babies, *Lancet,* i, 64, 1981.

232. **Wilkinson, A., Jenkins, P. A., and Jeffrey, J. A.,** Two controlled trials of dry artificial surfactant deficiency, *Lancet,* i, 287, 1985.

233. **Milner, A. D., Vyas, H., and Hopkin, I. E.,** Effects of artificial surfactant on lung function and blood gases in idiopathic respiratory distress syndrome, *Arch. Dis. Child.,* 58, 458, 1983.

234. **Durand, D. J., Clyman, R. I., Heymann, M. A., Clements, J. A., Mauray, F., Kitterman, J., and Ballard, P.,** Effects of a protein-free synthetic surfactant on survival and pulmonary function in preterm lambs, *J. Pediatr.,* 107, 775, 1985.

235. **Hallman, M., Merritt, T. A., Jarvenpaa, A. L., Boynton, B., Mannino, F., Gluck, L., Moore, T., and Edwards, D.,** Exogenous human surfactant for treatment of severe respiratory distress syndrome: a randomized prospective clinical trial, *J. Pediatr.,* 106, 963, 1985.

236. **Enhorning, G., Sheenan, A., Possmayer, F., Dunn, M., Chen, C. P., and Milligan, J.,** Prevention of neonatal respiratory distress syndrome by tracheal instillation of surfactant: a randomized clinical trial, *Pediatrics,* 76, 145, 1985.

237. **Kwong, M. S., Egan, E. A., Notter, R. H., and Shapiro, D. L.,** Double-blind clinical trial of calf lung surfactant extract for the prevention of Hyaline Membrane Disease in extremely premature infants, *Pediatrics,* 76, 585, 1985.

238. **Shapiro, D. L., Notter, R. H., Morin, F. C., Deluga, K. S., Golub, L. M., Sinkin, R. A., Weiss, K. I., and Cox, C.,** Double blind, randomized trial of calf lung surfactant administered at birth to very premature infants for the prevention of Respiratory Distress Syndrome, *Pediatrics,* 76, 593, 1985.

239. **Frosolono, M. F.,** Lung, in *Lipid Metabolism in Mammals,* Snyder, F., Ed., Plenum Press, New York, 1977, 1.

240. **Farrell, P. M.,** General features of phospholipid metabolism in the developing lung, in *Lung Development: Biological and Clinical Perspectives,* Vol. 1, Farrell, P. M., Ed., Academic Press, New York, 1982, 223.

241. **Bleasdale, J. E. and Johnston, J. M.,** Phosphatidic acid production and utilization, in *Lung Development: Biological and Clinical Perspectives,* Vol. 1, Farrell, P. M., Ed., Academic Press, New York, 1982, 259.

242. **Van Golde, L. M. G.,** Metabolism of phospholipids in the lung, *Am. Rev. Respir. Dis.,* 114, 977, 1976.

243. **Van Golde, L. M. G.,** Synthesis of surfactant lipids in the adult lung, *Am. Rev. Physiol.,* 47, 765, 1985.

244. **Batenburg, J. J. and van Golde, L. M. G.,** Formation of pulmonary surfactant in whole lung and isolated type II alveolar cells, *Rev. Perinat. Med.,* 3, 73, 1979.

245. **Batenburg, J. J.,** Biosynthesis and secretion of pulmonary surfactant, in *Pulmonary Surfactant,* Robertson, B., van Golde, L. M. G., and Batenburg, J. J., Eds., Elsevier, Amsterdam, 1984, 237.

246. **Sanders, R. L.,** Introduction to lipid biochemistry, in *Lung Development: Biological and Clinical Perspectives,* Vol. 1, Farrell, P. M., Ed., Academic Press, New York, 1982, 167.

247. **Vance, D. E. and Vance, J. E.,** *Biochemistry of Lipids and Membranes,* The Benjamin/Cummings Publ. Co., Menlo Park, Calif., 1985.

248. **van Golde, L. M. G. and van den Berg, S. C.,** Liver, in *Lipid Metabolism in Mammals,* Vol. 2, Snyder, F., Ed., Plenum Press, New York, 1977, 35.

249. **Brindley, D. N. and Sturton, R. G.,** Phosphatidate metabolism and its relation to triacylglycerol biosynthesis, in *Phospholipids,* Vol. 4, Hawthorne, J. N. and Ansell, G. B., Eds., Elsevier, Amsterdam, 1982, 179.

250. **Tamai, Y. and Lands, W. E. M.,** Positional specificity of *sn*-glycerol-3-phosphate acylation during phosphatidate formation by rat liver microsomes, *J. Biochem.,* 76, 847, 1974.

251. **Hendry, A. T. and Possmayer, F.,** Pulmonary phospholipid biosynthesis: properties of a stable microsomal glycerophosphate acyltransferase preparation from rabbit lung, *Biochim. Biophys. Acta,* 369, 156, 1974.

252. **Snyder, F. and Malone, B.,** Acyltransferases and the biosynthesis of pulmonary surfactant lipid in adenoma alveolar type II cells, *Biochem. Biophys. Res. Commun.,* 66, 914, 1975.

253. **van Golde, L. M. G., Den Breejen, J. N., and Batenburg, J. J.,** Isolated alveolar type II cells: a model for studies on the formation of surfactant dipalmitoylphosphatidylcholine, *Biochem. Soc. Trans.,* 13, 1087, 1985.

254. **Yamada, K. and Okuyama, H.,** Possible involvement of acyltransferase systems in the formation of pulmonary surfactant in the lung, *Arch. Biochem. Biophys.,* 196, 209, 1979.

255. **Finkelstein, J. N., Maniscalco, W. M., and Shapiro, D. L.,** Properties of freshly isolated type II alveolar epithelial cells, *Biochim. Biophys. Acta,* 762, 398, 1983.

256. **Chevalier, G. and Collet, A. J.,** In vivo incorporation of choline-³H, leucine-³H and galactose-³H in alveolar type II pneumocytes in relation to surfactant synthesis. A quantitative radioautographic study in mouse by electron microscopy, *Anat. Rec.,* 174, 289, 1972.

257. **Voelker, D. R. and Snyder, F.,** Subcellular site and mechanism of synthesis of disaturated phosphatidylcholine in alveolar type II cell adenomas, *J. Biol. Chem.,* 254, 8628, 1979.

258. **Das, S. K., McCullough, M. S., and Haldar, D.,** Acyl-CoA:*sn*-glycerol 3-phosphate acyltransferase in mitochondria and microsomes of adult and fetal guinea pig lung, *Biochem. Biophys. Res. Commun.,* 101, 237, 1981.

259. **Manning, R. and Brindley, D. N.,** Tritium isotope effects in the measurement of the glycerol phosphate and dihydroxyacetonephosphate pathways of glycerolipid biosynthesis in rat liver, *Biochem. J.,* 130, 1003, 1972.

260. **Mason, R. J.,** Importance of the acyldihydroxyacetone phosphate pathway in the synthesis of phosphatidylglycerol and phosphatidylcholine in alveolar type II cells, *J. Biol. Chem.,* 253, 3367, 1978.

261. **Ide, H. and Weinhold, P. A.,** Properties of diacylglycerol kinase in adult and fetal rat lung, *Biochim. Biophys. Acta,* 713, 547, 1982.

262. **Berridge, M. J. and Irvine, R. F.,** Inositol trisphosphate, a novel second messenger in cellular signal transduction, *Nature (London),* 312, 315, 1984.

263. **Sano, K., Voelker, D. R., and Mason, R. J.,** Involvement of protein kinase C in pulmonary surfactant secretion from alveolar type II cells, *J. Biol. Chem.,* 260, 12725, 1985.

264. **Van Golde, L. M. G.,** The CDP choline pathway: cholinephosphotransferase, in *Lung Development: Biological and Clinical Perspectives,* Vol. 1, Farrell, P. M., Ed., Academic Press, New York, 1982, 337.

265. **Stith, I. E. and Das, S. K.,** Development of cholinephosphotransferase in guinea pig lung mitochondria and microsomes, *Biochim. Biophys. Acta,* 714, 250, 1982.

266. **Miller, J. C. and Weinhold, P. A.,** Cholinephosphotransferase in rat lung. The in vitro synthesis of dipalmitoylcholine from dipalmitoylglycerol, *J. Biol. Chem.,* 256, 12662, 1981.

267. **Yost, R. W., Chander, A., Dodia, C., and Fisher, A. B.,** Stimulation of the methylation pathway for phosphatidylcholine synthesis in rat lungs by choline deficiency, *Biochim. Biophys. Acta,* 875, 738, 1986.

268. **Sahu, S. and Lynn, W.,** Characterization of phospholipase A from pulmonary secretions of patients with alveolar proteinosis, *Biochim. Biophys. Acta,* 489, 307, 1977.

269. **Ohta, M., Hasegawa, H., and Ohno, K.,** Calcium independent phospholipase A₂ activity in rat lung supernatant, *Biochim. Biophys. Acta,* 280, 552, 1972.

270. **Hallman, M., Spragg, R., Harwell, J. H., Moser, K. M., and Gluck, L.,** Evidence of lung failure abnormality in respiratory failure: study of bronchoalveolar lavage phospholipids, surface activity, phospholipase activity and plasma myoinositol, *J. Clin. Invest.,* 70, 673, 1982.

271. **Franson, R. C. and Weir, D. L.,** Isolation and characterization of a membrane-associated calcium-dependent phospholipase A₂ from rabbit lung, *Lung,* 160, 275, 1982.

272. **Melin, B., Maximilien, R., Friedlander, G., Etienne, J., and Alcindor, L. G.,** Activites phospholipasiques pulmonaires du foetus de rat: variations au cours du developpement, *Biochim. Biophys. Acta,* 486, 590, 1977.

273. **Longmore, W. J., Oldenburg, U., and van Golde, L. M. G.,** Phospholipase A₂ in rat lung microsomes: substrate specificity towards endogenous phosphatidylcholines, *Biochim. Biophys. Acta,* 572, 452, 1979.

274. **Heath, M. F. and Jacobson, W.,** Phospholipases A₁ and A₂ in lamellar inclusion bodies of the alveolar epithelium of rabbit lung, *Biochim. Biophys. Acta,* 441, 443, 1976.

275. **Rao, R. H., Waite, M., and Myrvik, Q. N.,** Deacylation of dipalmitoyl lecithin by phospholipase A in alveolar macrophages, *Exp. Lung Res.,* 2, 9, 1981.

276. **Akino, T., Abe, M., and Arai, T.,** Studies on the biosynthetic pathways of molecular species of lecithin by rat lung slices, *Biochim. Biophys. Acta,* 248, 274, 1971.

277. **Erbland, J. F. and Marinetti, G. V.,** The enzymatic acylation and hydrolysis of lysolecithin, *Biochim. Biophys. Acta,* 106, 128, 1965.

278. **Van Heusden, G. P. H., Noteborn, J. M., and van den Bosch, H.,** Selective utilization of palmitoyl lysophosphatidylcholine in the synthesis of disaturated phosphatidylcholine in rat lung. A combined in vitro and in vivo approach, *Biochim. Biophys. Acta,* 664, 49, 1981.

279. **Vianen, G. M. and van den Bosch, H.,** Lysophospholipase and lysophosphatidylcholine: lysophosphatidylcholine transacylase from rat lung: evidence for a single enzyme and some aspects of its specificity, *Arch. Biochem. Biophys.,* 190, 373, 1978.

280. **Van Heusden, G. P. H., Reutelingsperger, C. P. M., and van den Bosch, H.,** Substrate specificity of lysophospholipase-transacylase from rat lung and its action on various physical forms of lysophosphatidylcholine, *Biochim. Biophys. Acta,* 663, 22, 1981.

281. **Van Heusden, G. P. H., Vianen, G. M., and van den Bosch, H.,** Differentiation between acyl CoA:lysophosphatidylcholine acyltransferase and lysophosphatidylcholine:lysophosphatidylcholine transacylase in the synthesis of dipalmitoylphosphatidylcholine in rat lung, *J. Biol. Chem.,* 255, 9312, 1980.

282. **de Vries, A. C. J., Batenburg, J. J., and Van Golde, L. M. G.,** Lysophosphatidylcholine acyltransferase and lysophosphatidylcholine:lysophosphatidylcholine acyltransferase in alveolar type II cells from fetal rat lung, *Biochim. Biophys. Acta,* 833, 93, 1985.

283. **Batenburg, J. J., Longmore, W. J., Klazinga, W., and van Golde, L. M. G.,** Lysolecithin acyltransferase and lysolecithin:lysolecithin acyltransferase in adult lung alveolar type II epithelial cells, *Biochim. Biophys. Acta,* 573, 136, 1979.

284. **Crecelius, C. A. and Longmore, W. J.,** Acyltransferase activities in adult rat type II pneumonocyte-derived subcellular fractions, *Biochim. Biophys. Acta,* 295, 238, 1984.

285. **Van Heusden, G. P. H. and van den Bosch, H.,** Comparison of acyl-CoA:lysophosphatidylcholine acyltransferase and lysophosphatidylcholine:lysophosphatidylcholine transacylase activity in various mammalian lungs, *Biochim. Biophys. Acta,* 666, 508, 1981.

286. **Chan, F., Harding, P. G. R., Wong, T., Fellows, G. F., and Possmayer, F.,** Cellular distribution of enzymes involved in phosphatidylcholine synthesis in developing rat lung, *Can. J. Biochem. Cell Biol.,* 61, 107, 1983.

287. **Hill, E. E. and Lands, W. E. M.,** Phospholipid metabolism, in *Lipid Metabolism,* Wakil, S. J., Ed., Academic Press, New York, 1970, 185.

288. **Hasegawa, H. and Ohno, K.,** Acyl-CoA: 1-acyl-lysolecithin acyltransferase activities in rat lung microsomes, *Biochim. Biophys. Acta,* 380, 486, 1975.

289. **Holub, B. J. and Piekarski, J.,** Relative suitability of 1-palmitoyl and 1-stearoyl homologues of 1-acyl-*sn*-glycerylphosphoryl choline and different acyl donors for phosphatidylcholine synthesis via acyl-CoA:1-acyl-*sn*-glycero-3-phosphorylcholine acyltransferase in rat lung microsomes, *Can. J. Biochem.,* 58, 434, 1980.

290. **Okuyama, H., Yamada, K., Miyagawa, T., Suzuki, M., Prasad, R., and Lands, W. E. M.,** Enzymatic basis for the formation of pulmonary surfactant lipids by acyltransferase systems, *Arch. Biochem. Biophys.,* 221, 99, 1983.

291. **Ide, H. and Weinhold, P. A.,** Cholinephosphotransferase in rat lung: in vitro formation of dipalmitoyl phosphatidylcholine and general lack of specificity using endogenously generated diacylglycerol, *J. Biol. Chem.,* 257, 14926, 1981.

292. **Batenburg, J. J., Post, M., Oldenborg, V., and Van Golde, L. M. G.,** The perfused isolated lung as a possible model for the study of lipid synthesis by type II cells in their natural environment, *Exp. Lung Res.,* 1, 57, 1980.

293. **Post, M., Schuurmans, E. A. J. M., Batenburg, J. J., and Van Golde, L. M. G.,** Mechanisms involved in the synthesis of disaturated phosphatidylcholine by alveolar type II cells isolated from adult rat lung, *Biochim. Biophys. Acta,* 750, 68, 1983.

294. **Mason, R. J. and Dobbs, L. G.,** Synthesis of phosphatidylcholine and phosphatidylglycerol by alveolar type II cells in primary culture, *J. Biol. Chem.,* 225, 5101, 1980.

295. **Mason, R. J. and Nellenbogen, J.,** Synthesis of saturated phosphatidylcholine and phosphatidylglycerol by freshly isolated rat alveolar type II cells, *Biochim. Biophys. Acta,* 794, 392, 1984.

296. **Batenburg, J. J., Longmore, W. J., and Van Golde, L. M. G.,** The synthesis of phosphatidylcholine by adult rat lung alveolar type II epithelial cells in primary culture, *Biochim. Biophys. Acta,* 529, 160, 1978.

297. **Buechler, K. F. and Rhoades, R.,** De novo fatty acid synthesis in the perfused rat lung, *Biochim. Biophys. Acta,* 665, 393, 1981.

298. **Engle, M. J., Sanders, R. L., and Longmore, W. J.,** Evidence for the synthesis of lung surfactant dipalmitoyl phosphatidylcholine by a ''remodeling'' mechanism, *Biochem. Biophys. Res. Commun.,* 94, 23, 1980.

299. **Kyei-Aboage, K., Rubinstein, D., and Beck, J. D.,** Biosynthesis of dipalmitol lecithin by the rabbit lung, *Can. J. Biochem.,* 51, 1581, 1973.

300. **Nijssen, J. G. and van den Bosch, H.,** Cytosol-stimulated remodeling of phosphatidylcholine in rat lung microsomes, *Biochim. Biophys. Acta,* 875, 450, 1986.

301. **Nijssen, J. G. and van den Bosch, H.,** Coenzyme A-mediated transacylation of *sn*-2 fatty acids from phosphatidylcholine in rat lung microsomes, *Biochim. Biophys. Acta,* 875, 458, 1986.

302. **Stymne, S. and Stobart, A. K.,** Involvement of acyl exchange between acyl-CoA and phosphatidylcholine in the remodelling of phosphatidylcholine in microsomal preparations of rat lung, *Biochim. Biophys. Acta,* 837, 239, 1985.

303. **Moriya, T. and Kanoh, H.,** In vivo studies on the de novo synthesis of molecular species of rat lung lecithins, *Tohoku J. Exp. Med.,* 112, 241, 1974.

304. **Van Heusden, G. P. H. and van den Bosch, H.,** Utilization of disaturated and unsaturated phosphatidylcholine and diacylglycerols by cholinephosphotransferase in rat lung microsomes, *Biochim. Biophys. Acta,* 711, 361, 1982.

305. **Wykle, R. L., Malone, B., and Snyder, F.,** Biosynthesis of dipalmitoyl-*sn*-glycero-3-phosphocholine by adenoma alveolar type II cells, *Arch. Biochem. Biophys.,* 181, 249, 1977.

306. **Creceluis, C. A. and Longmore, W. J.,** A study of the molecular species of diacylglycerol, phosphatidylcholine and phosphatidylethanolamine and of cholinephosphotransferase and ethanolaminephosphotransferase activities in the type II pneumocyte, *Biochim. Biophys. Acta,* 795, 247, 1984.

307. **Hallman, M., Saugstad, O. D., Porreco, R. R., Epstein, B. L., and Gluck, L.,** Role of myoinositol in regulation of surfactant phospholipids in the newborn, *Early Human Dev.,* 10, 245, 1985.

308. **Beppu, O. S., Clements, J. A., and Goerke, J.,** Phosphatidylglycerol-deficient lung surfactant has normal properties, *J. Appl. Physiol.,* 55, 496, 1983.

309. **Hallman, M., Enhorning, G., and Possmayer, F.,** Composition and surface activity of normal and phosphatidylglycerol-deficient lung surfactant, *Pediatr. Res.,* 19, 286, 1985.

310. **Hallman, M. and Gluck, L.,** Formation of acidic phospholipids in rabbit lung during perinatal development, *Pediatr. Res.,* 14, 1250, 1980.

311. **Longmuir, K. J. and Johnston, J. M.,** Changes in CTP:phosphatidate cytidylyltransferase activity during rabbit lung development, *Biochim. Biophys. Acta,* 620, 500, 1980.

312. **Harding, P. G. R., Chan, F., Casola, P. G., Fellows, G. F., Wong, T., and Possmayer, F.,** Subcellular distribution of the enzymes related to phospholipid synthesis in developing rat lung, *Biochim. Biophys. Acta,* 750, 373, 1983.

313. **Bleasdale, J. E., Wallis, P., MacDonald, P. C., and Johnston, J. M.,** Characterization of the forward and reverse reactions catalysed by CDPdiacylglycerol:inositol transferase in rabbit lung tissue, *Biochim. Biophys. Acta,* 575, 135, 1979.

314. **Bleasdale, J. E., Wallis, P., MacDonald, P. C., and Johnston, J. M.,** Changes in CDP-diglyceride inositol transferase activity during rabbit lung development, *Pediatr. Res.,* 13, 1182, 1979.

315. **Bleasdale, J. E. and Wallis, P.,** Phosphatidylinositol-inositol exchange in rabbit lung, *Biochim. Biophys. Acta,* 664, 428, 1981.

316. **Bleasdale, J. E. and Johnston, J. M.,** CMP-dependent incorporation of [^{14}C]glycerol 3-phosphate into phosphatidylglycerol and phosphatidylglycerophosphate by rabbit lung microsomes, *Biochim. Biophys. Acta,* 710, 377, 1982.

317. **Mavis, R. D. and Vang, M. J.,** Optimal assay and subcellular location of phosphatidylglycerol synthesis in lung, *Biochim. Biophys. Acta,* 664, 409, 1981.

318. **Hallman, M. and Gluck, L.,** Phosphatidylglycerol in lung surfactant. II. Subcellular distribution and mechanism of biosynthesis in vitro, *Biochim. Biophys. Acta,* 409, 172, 1975.

319. **Bleasdale, J. E., Tyler, N. E., and Snyder, J. M.,** Subcellular sites of synthesis of phosphatidylglycerol and phosphatidylinositol in type II pneumonocytes, *Lung,* 165, 345, 1986.

320. **Batenburg, J. J., Klazinga, W., and van Golde, L. M. G.,** Regulation and location of phosphatidylglycerol and phosphatidylinositol synthesis in type II cells isolated from fetal rat lung, *Biochim. Biophys. Acta,* 833, 17, 1985.

321. **Hallman, M. and Epstein, B. L.,** Role of myoinositol in the synthesis of phosphatidylglycerol and phosphatidylinositol in the lung, *Biochem. Biophys. Res. Commun.,* 92, 1151, 1980.

322. **Hallman, M., Slivka, S., Wozniak, P., and Sils, J.,** Perinatal development of myoinositol uptake into lung cells: surfactant phosphatidylglycerol and phosphatidylinositol synthesis in the rabbit, *Pediatr. Res.,* 20, 179, 1986.

323. **Esko, J. D. and Raetz, C. R. H.,** Mutants of Chinese hamster ovary cells with altered membrane phospholipid composition. Replacement of phosphatidylinositol by phosphatidylglycerol in a myo-inositol auxotroph, *J. Biol. Chem.,* 255, 4474, 1980.

324. **Kates, M.,** Hydrolysis of lecithin by plant plastid enzymes, *Can. J. Biochem.,* 35, 575, 1955.

325. **Agranoff, B. W.,** Hydrolysis of long-chain alkyl phosphates and phosphatidic acid by an enzyme purified from pig brain, *J. Lipid Res.,* 3, 190, 1962.

326. **Johnston, J. M. and Bearden, J. H.,** Intestinal phosphatidate phosphatase, *Biochim. Biophys. Acta,* 56, 365, 1962.

327. **Coleman, R. and Hübscher, G.,** Metabolism of phospholipids. V. Studies on phosphatidic acid phosphatase, *Biochim. Biophys. Acta,* 56, 479, 1962.

328. **Wilgram, G. F. and Kennedy, E. P.,** Intracellular distribution of some enzymes catalyzing reactions in the biosynthesis of complex lipids, *J. Biol. Chem.,* 238, 2615, 1963.

329. **Sedgwick, B. and Hübscher, G.,** Metabolism of phospholipids. IX. Phosphatidate phosphohydrolase in rat liver, *Biochim. Biophys. Acta,* 196, 63, 1965.

330. **Hokin, L. E., Hokin, M. R., and Mathison, D.,** Phosphatidic acid phosphatase in erythrocyte membrane, *Biochim. Biophys. Acta,* 67, 485, 1963.

331. **Schultz, F. M., Jimenez, J. M., MacDonald, P. C., and Johnston, J. M.,** Fetal lung maturation. I. Phosphatidic acid phosphohydrolase in rabbit lung, *Gynecol. Invest.,* 5, 222, 1974.

332. **Jimenez, J. M., Schultz, F. M., MacDonald, P. C., and Johnston, J. M.,** Fetal lung maturation. II. Phosphatidic acid phosphohydrolase in human amniotic fluid, *Gynecol. Invest.,* 5, 245, 1984.

333. **Ravinuthula, H. R., Miller, J. C., and Weinhold, P. A.,** Phosphatidate phosphohydrolase. Activity and properties in fetal and adult rat lung, *Biochim. Biophys. Acta,* 530, 347, 1978.

334. **Mavis, R. D., Finkelstein, J. N., and Hall, B. P.**, Pulmonary surfactant synthesis. A highly active microsomal phosphatidate phosphohydrolase in the lung, *J. Lipid Res.*, 19, 467, 1978.

335. **Yeung, A., Casola, P. G., Wong, C., Fellows, J. F., and Possmayer, F.**, Pulmonary phosphatidic acid phosphatase: a comparative study of the aqueously-dispersed phosphatidate-dependent and the membrane-bound phosphatidate-dependent phosphatidic acid phosphatase activities of rat lung, *Biochim. Biophys. Acta*, 574, 226, 1970.

336. **Spitzer, H. L. and Johnston, J. M.**, Characterization of phosphatidate phosphohydrolase activity associated with isolated lamellar bodies, *Biochim. Biophys. Acta*, 531, 275, 1978.

337. **Jimenez, J. M. and Johnston, J. M.**, The release of phosphatidic acid phosphohydrolase and phospholipids into the human amniotic fluid, *Pediatr. Res.*, 10, 767, 1976.

338. **Herbert, W. P., Johnston, J. M., McDonald, P. C., and Jimenez, J. M.**, Human amniotic fluid phosphatidate phosphohydrolase activity through normal gestation and its relation to the lecithin/sphingomyelin ratio, *Am. J. Obstet. Gynecol.*, 132, 373, 1978.

339. **Spitzer, H. L., Rice, R. M., MacDonald, P. C., and Johnston, J. M.**, Phospholipid biosynthesis in lung lamellar bodies, *Biochem. Biophys. Res. Commun.*, 66, 17, 1975.

340. **Johnston, J. M., Reynolds, G., Wylie, M. B., and MacDonald, P. C.**, The phosphohydrolase activity in lamellar bodies and its relationship to phosphatidylglycerol and lung surfactant formation, *Biochim. Biophys. Acta*, 531, 65, 1978.

341. **Delahunty, T. J., Spitzer, H. L., Jimenez, J. M., and Johnston, J. M.**, Phosphatidate phosphohydrolase activity in porcine pulmonary surfactant, *Am. Rev. Respir. Dis.*, 119, 75, 1979.

342. **Okazaki, T. and Johnston, J. M.**, Distribution of the phosphatidate phosphohydrolase activity in the lamellar body and lysosomal fractions, *Lipids*, 15, 447, 1980.

343. **Rosenfeld, C. R., Andujo, O., Johnston, J. M., and Jimenez, J. M.**, Phosphatidic acid phosphohydrolase and phospholipids in tracheal and amniotic fluids during normal ovine pregnancy, *Pediatr. Res.*, 14, 891, 1980.

344. **Okazaki, T., Johnston, J. M., and Snyder, J. M.**, Morphogenesis of the lamellar body in fetal lung tissue in vitro, *Biochim. Biophys. Acta*, 712, 283, 1982.

345. **McMurray, W. C. and Dawson, R. M. C.**, Phospholipid exchange reactions within the liver cells, *Biochem. J.*, 112, 91, 1969.

346. **Benson, B. M.**, Properties of an acid phosphatase in pulmonary surfactant, *Proc. Natl. Acad. Sci. U.S.A.*, 77, 808, 1980.

347. **Casola, P. G., MacDonald, P. M., McMurray, W. C., and Possmayer, F.**, Concerning the coidentity of phosphatidic acid phosphohydrolase and phosphatidyl glycerol phosphohydrolase in rat lung lamellar bodies, *Exp. Lung Res.*, 3, 1, 1982.

348. **Casola, P. G. and Possmayer, F.**, Pulmonary phosphatidic acid phosphohydrolase. Developmental patterns in rabbit lung, *Biochim. Biophys. Acta*, 665, 186, 1981.

349. **Rooney, S. A., Gorban, L. I., Marino, P. A., Maniscalco, W. M., and Gross, I.**, Effects of betamethasone on phospholipid content, composition and biosynthesis in the fetal rabbit lung, *Biochim. Biophys. Acta*, 572, 64, 1979.

350. **Brehier, A., Benson, B. J., Williams, M. C., Mason, R. J., and Ballard, P. L.**, Corticosteroid induction of phosphatidic acid phosphatase in fetal rabbit lung, *Biochem. Biophys. Res. Commun.*, 77, 883, 1977.

351. **Possmayer, F., Duwe, G., Metcalfe, R., Stewart-Dehaan, P. J., Wong, C., Las Heras, J., and Harding, P. G. R.**, Cortisol induction of pulmonary maturation in the rabbit foetus, *Biochem. J.*, 166, 487, 1977.

352. **Possmayer, F., Casola, P., Chan, F., Hill, S., Metcalfe, I. R., Stewart-DeHaan, P. J., Wong, T., Las Heras, J., Gammal, E. G., and Harding, P. G. R.**, Glucocorticoid induction of pulmonary maturation in the rabbit fetus. The effect of maternal injection of betamethasone on the activity of enzymes in fetal lung, *Biochim. Biophys. Acta*, 574, 197, 1979.

353. **Freese, W. B. and Hallman, M.**, The effect of betamethasone and fetal sex on the synthesis and maturation of lung surfactant phospholipids in rabbits, *Biochim. Biophys. Acta*, 750, 47, 1983.

354. **Khosla, S., Gobran, L. I., and Rooney, S. A.**, Stimulation of phosphatidylcholine synthesis by 17beta-estradiol in fetal rabbit lung, *Biochim. Biophys. Acta*, 617, 282, 1980.

355. **Possmayer, F., Casola, P. G., Chan, F., MacDonald, P., Ormseth, M. A., Wong, T., Harding, P. G. R., and Tokmakjian, S.**, Hormonal induction of pulmonary maturation in the rabbit fetus. Effects of treatment with estradiol-17beta on the endogenous levels of cholinephosphate, CDP-choline and phosphatidylcholine, *Biochim. Biophys. Acta*, 664, 10, 1981.

356. **Maniscalco, W. M., Wilson, C. M., Gross, I., Gobran, L., Rooney, S. A., and Warshaw, J. B.**, Development of glycogen and phospholipid metabolism in fetal and newborn rat lung, *Biochim. Biophys. Acta*, 530, 333, 1978.

357. **Casola, P. G. and Possmayer, F.**, Pulmonary phosphatidic acid phosphohydrolase. Developmental patterns in rat lung, *Biochim. Biophys. Acta*, 665, 177, 1981.

358. **Filler, D. A. and Rhoades, R. A.**, Lung phosphatidate phosphatase activity during altered physiological states, *Exp. Lung Res.*, 3, 37, 1979.

359. **Crecelius, C. A. and Longmore, W. J.,** Phosphatidic acid phosphatase activity in subcellular fractions derived from adult rat type II pneumonocytes in primary culture, *Biochim. Biophys. Acta,* 750, 447, 1983.
360. **Tietz, D. F. and Shapiro, B.,** The synthesis of glycerides in liver homogenates, *Biochim. Biophys. Acta,* 19, 374, 1956.
361. **Stein, Y., Tietz, A., and Shapiro, B.,** Glyceride synthesis by rat liver mitochondria, *Biochim. Biophys. Acta,* 26, 286, 1957.
362. **Smith, M. E. and Hübscher, G.,** The biosynthesis of glycerides by mitochondria from rat liver. The requirements for a soluble protein, *Biochem. J.,* 101, 308, 1966.
363. **Hübscher, G., Brindley, D. N., Smith, M. E., and Sedgwick, B.,** Stimulation of biosynthesis of glycerides, *Nature (London),* 216, 449, 1967.
364. **Johnston, J. M., Rao, G. A., Lowe, P. A., and Schwarz, B. E.,** The nature of the stimulatory role of the supernatant fraction on triglyceride synthesis by the alpha-glycerophosphate pathway, *Lipids,* 2, 14, 1967.
365. **Smith, M. E., Sedgwick, B., Brindley, D. N., and Hübscher, G.,** The role of phosphatidate phosphohydrolase in glyceride biosynthesis, *Eur. J. Biochem.,* 3, 70, 1967.
366. **Mitchell, M. P., Brindley, D. N., and Hübscher, G.,** Properties of phosphatidate phosphohydrolase, *Eur. J. Biochem.,* 18, 214, 1971.
367. **Brindley, D. N.,** Intracellular translocation of phosphatidate phosphohydrolase and its possible role in the control of glycerolipid synthesis, *Prog. Lipid Res.,* 23, 115, 1984.
368. **Gatt, S.,** Inhibitors of enzymes of phospholipid and sphingomyelin metabolism, in *Metabolic Inhibitors,* Vol. 3, Questl, G. H. and Kates, M., Eds., Academic Press, New York, 1972, 349.
369. **Gatt, S. and Barenholz, Y.,** Enzymes of complex lipid metabolism, *Annu. Rev. Biochem.,* 42, 61, 1973.
370. **Brindley, D. N. and White, D. A.,** Difficulties encountered in interpreting the kinetics of enzyme reactions involving lipid substrates, *Biochem. Soc. Trans.,* 2, 44, 1974.
371. **Fallon, H. J., Barwick, J., Lamb, R. G., and van den Bosch, H.,** Studies of rat liver microsomal diacylglycerol acyltransferase and cholinephosphotransferase using microsomal-bound substrate: effects of high fructose intake, *J. Lipid Res.,* 16, 107, 1975.
372. **Fallon, H. J., Lamb, J. R., and Jambdar, S. C.,** Phosphatidate phosphohydrolase and the regulation of glycerolipid biosynthesis, *Biochem. Soc. Trans.,* 5, 37, 1977.
373. **Coleman, R. and Bell, R. M.,** Phospholipid synthesis in isolated fat cells. Studies on microsomal diacylglycerol cholinephosphotransferase and diacylglycerol ethanolaminephosphotransferase activities, *J. Biol. Chem.,* 252, 3050, 1977.
374. **Kanoh, H. and Ohno, K.,** Substrate-selectivity of rat liver microsomal 1,2-diacylglycerol:CDP choline (ethanolamine) choline (ethanolamine) phosphotransferase in utilizing endogenous substrates, *Biochim. Biophys. Acta,* 380, 199, 1975.
375. **Sarzala, M. G. and Van Golde, L. M. G.,** Selective utilization of endogenous unsaturated phosphatidylcholines and diacylglycerols by cholinephosphotransferase of mouse lung microsomes, *Biochim. Biophys. Acta,* 441, 423, 1976.
376. **Holub, B. and Piekarski, J.,** Biosynthesis of molecular species of CDP-diacylglycerol from endogenously-labelled phosphatidate in rat liver microsomes, *Lipids,* 11, 251, 1976.
377. **Van Heusden, G. P. H. and van den Bosch, H.,** The influence of exogenous and membrane-bound phosphatidate concentration on the activity of CTP:phosphatidate cytidylyltransferase and phosphatidate phosphohydrolase, *Eur. J. Biochem.,* 84, 405, 1978.
378. **Casola, P. G. and Possmayer, F.,** Pulmonary phosphatidic acid phosphohydrolase. Properties of membrane-bound phosphatidate-dependent phosphatidic acid phosphatase in rat lung, *Biochim. Biophys. Acta,* 574, 212, 1979.
379. **Casola, P. G. and Possmayer, F.,** Pulmonary phosphatidic acid phosphohydrolase: further studies on the activities in rat lung responsible for the hydrolysis of membrane-bound and aqueously-dispersed phosphatidate, *Can. J. Biochem.,* 59, 500, 1981.
380. **Casola, P. G. and Possmayer, F.,** Separation and characterization of the membrane-bound and aqueously-dispersed phosphatidate phosphatidic acid phosphohydrolase activities in rat lung, *Biochim. Biophys. Acta,* 664, 298, 1981.
381. **Sturton, R. G. and Brindley, D. N.,** Problems encountered in measuring the activity of phosphatidate phosphohydrolase, *Biochem. J.,* 171, 263, 1978.
382. **Davis, C. S. G.,** Purification and Characterization of Phosphatidate Phosphohydrolase from a Soluble Fraction of Brain, Ph.D. thesis, University of Michigan, Ann Arbor, Mich., 1976.
383. **Jamdar, S. C. and Fallon, H. J.,** Glycerolipid synthesis in rat adipose tissue. II. Properties and distribution of phosphatidate phosphatase, *J. Lipid Res.,* 14, 517, 1973.
384. **Moller, F., Green, P., and Harkness, E. J.,** Soluble rat phosphatidate phosphohydrolase: characterization and effect of fasting and various lipids, *Biochim. Biophys. Acta,* 486, 359, 1977.
385. **Rooney, S. A. and Gobran, L. I.,** Alveolar wash and lavaged lung tissue phosphatidylcholine composition during fetal rabbit development, *Lipids,* 12, 1050, 1977.

386. **Rooney, S. A., Gobran, L. I., and Wai-Lee, T. S.,** Stimulation of surfactant production by oxytocin-induced labour in the rabbit, *J. Clin. Invest.,* 60, 754, 1977.

387. **Walton, P. A. and Possmayer, F.,** Mg^{2+}-dependent phosphatidate phosphohydrolase of rat lung. Development of an assay employing a defined chemical substrate which reflects the phosphohydrolase activity measured using membrane-bound substrate, *Anal. Biochem.,* 151, 479, 1985.

388. **Flynn, T. J., Deshmukd, D. S., and Pieringer, R. A.,** A rapid and sensitive radiochemical assay for phosphatidate phosphohydrolase activity, *J. Lipid Res.,* 18, 777, 1977.

389. **Sturton, R. G., Pritchard, P. H., Han, L.-Y., and Brindley, D. N.,** The involvement of phosphatidate phosphohydrolase and phospholipase A activities in the control of hepatic glycerolipid synthesis, *Biochem. J.,* 174, 667, 1978.

390. **Sturton, R. G. and Brindley, D. N.,** Factors controlling the metabolism of phosphatidate by phosphohydrolase and phospholipase A-type activities. Effects of magnesium, calcium, and amphiphilic cationic drugs, *Biochim. Biophys. Acta,* 619, 494, 1980.

391. **Douglas, W. H. J., Sommers-Smith, S. K., and Johnston, J. M.,** Phosphatidate phosphohydrolase activity as a marker for surfactant synthesis in organotypic cultures of Type II alveolar pneumocytes, *J. Cell. Sci.,* 60, 199, 1983.

392. **Bowley, M., Cooling, J., Burditt, S. L., and Brindley, D. N.,** The effects of amphiphilic cationic drugs and inorganic cations on the activity of phosphatidate phosphohydrolase, *Biochem. J.,* 165, 447, 1977.

393. **Mangiapane, E. H., Lloyd-Davies, K. A., and Brindley, D. N.,** A study of some enzymes of glycerolipid biosynthesis in rat liver after subtotal hepatectomy, *Biochem. J.,* 134, 103, 1973.

394. **Lehtonen, M. A., Savolainen, M. J., and Hassinen, I. E.,** Hormone regulation of hepatic soluble phosphatidate phosphohydrolase: induction by cortisol in vivo and in isolated perfused rat liver, *FEBS Lett.,* 99, 162, 1979.

395. **Lamb, R. G. and Fallon, H. J.,** Glycerolipid formation from sn-glycerol-3-phosphate by rat liver cell fractions. The role of phosphatidate phosphohydrolase, *Biochim. Biophys. Acta,* 348, 166, 1974.

396. **Jamdar, S. C., Shapiro, D., and Fallon, H. J.,** Triacylglycerol biosynthesis in the adipose tissue of the obese-hyperglycaemic mouse, *Biochem. J.,* 158, 327, 1976.

397. **Whiting, P. H., Bowley, M., Sturton, R. G., Pritchard, P. H., Brindley, D. N., and Hawthorne, J. N.,** The effect of chronic diabetes, induced by streptozotocin, on the activities of some enzymes of glycerolipid synthesis in rat liver, *Biochem. J.,* 168, 147, 1977.

398. **Savolainen, M. J.,** Stimulation of hepatic phosphatidate phosphohydrolase activity by a single dose of ethanol, *Biochem. Biophys. Res. Commun.,* 75, 511, 1977.

399. **Cheng, C. H. K. and Saggerson, E. D.,** The inactivation of rat adipocyte Mg^{2+}-dependent phosphatidate phosphohydrolase by noradrenaline, *Biochem. J.,* 190, 659, 1980.

400. **Brindley, D. N., Cooling, J., Burdett, S. L., Pritchard, P. H., Pawson, S., and Sturton, R. G.,** The involvement of glucocorticoids in regulating the activity of phosphatidate phosphohydrolase and the synthesis of triacylglycerols in the liver, *Biochem. J.,* 180, 195, 1979.

401. **Germershausen, J. I., Yudkovitz, J. B., and Greenspan, M. D.,** A sensitive assay for phosphatidate phosphohydrolase in mouse liver microsomes, *Biochim. Biophys. Acta,* 620, 562, 1980.

402. **Walton, P. A. and Possmayer, F.,** The role of Mg^{2+}-dependent phosphatidate phosphohydrolase in pulmonary glycerolipid biosynthesis, *Biochim. Biophys. Acta,* 796, 364, 1984.

403. **Possmayer, F., Walton, P. A., and Harding, P. G. R.,** The relationship between the Mg-dependent and Mg-independent phosphatidic acid phosphohydrolases and glycerolipid biosynthesis in rat lung microsomes, in *The Physiologic Development of the Fetus and Newborn,* Jones, C. T. and Nathanielsz, P. W., Eds., Academic Press, New York, 1985, 235.

404. **Roncari, D. A. K. and Mack, E. Y. W.,** Purification of liver cytosolic proteins that stimulate triacylglycerol synthesis, *Can. J. Biochem.,* 59, 944, 1981.

405. **Roncari, D. A. K. and Mack, Y. W.,** Stimulation of triglyceride synthesis in mammalian liver and adipose tissue by two cytosolic compounds, *Biochem. Biophys. Res. Commun.,* 67, 790, 1975.

406. **Farrell, P. M. and Zachman, R. D.,** Enhancement of lecithin synthesis and phosphorylcholine glyceride transferase activity in the fetal rabbit lung after corticosteroid administration, *Pediatr. Res.,* 6, 337, 1972.

407. **Farrell, P. M., Blackburn, W. R., and Adams, A. J.,** Lung phosphatidylcholine synthesis and cholinephosphotransferase activity in anencephalic rat fetuses with corticosteroid deficiency, *Pediatr. Res.,* 11, 770, 1977.

408. **Rooney, S. A., Gobran, L., Gross, I., Wai-Lee, T. S., Nardone, L. L., and Motoyama, E. K.,** Studies on pulmonary surfactant. Effects of cortisol administration to fetal rabbits on lung phospholipid content, composition and biosynthesis, *Biochim. Biophys. Acta,* 450, 121, 1976.

409. **Gilfillan, A. M., Smart, D. A., and Rooney, S. A.,** Phosphatidylgylcerol stimulates cholinephosphate cytidylyltransferase activity and phosphatidylcholine synthesis in type II pneumonocytes, *Biochim. Biophys. Acta,* 835, 141, 1985.

410. **Walton, P. A. and Possmayer, F.,** Translocation of Mg^{2+}-dependent phosphatidate phosphohydrolase between cytosol and endoplasmic reticulum in a permanent cell line from human lung, *J. Biochem. Cell Biol.,* 64, 1135, 1986.

411. **Vance, D. E. and Pelech, S. L.,** Enzyme translocation in the regulation of phosphatidylcholine biosynthesis, *Trends Biochem. Sci.,* 9, 17, 1984.

412. **Vance, D. E. and Pelech, S. L.,** Cellular translocation of CTP:-phosphocholine cytidylyltransferase regulates the synthesis of CDP-choline, in *Novel Biochemical, Pharmacological and Clinical Aspects of Cytidinediphosphocholine,* Zappia, V., Kennedy, E. P., Nilsson, B. I., and Galletti, P., Eds., Elsevier, Amsterdam, Amsterdam, 1985, 15.

413. **Pelech, S. L. and Vance, D. E.,** Regulation of phosphatidylcholine biosynthesis, *Biochim. Biophys. Acta,* 779, 217, 1984.

414. **Possmayer, F.,** Pool sizes of the precursors for phosphatidylcholine synthesis in developing lung, in *Novel Biochemical, Pharmacological and Clinical Aspects of CDP-Choline,* Zappia, V., Kennedy, E. P., Nilsson, B. I., and Galletti, P., Eds., Elsevier, Amsterdam, 1985, 91.

415. **Rooney, S. A.,** Biosynthesis of lung surfactant during fetal and postnatal development, *Trends Biochem. Sci.,* 4, 189, 1979.

416. **Rooney, S. A. and Brehier, A.,** The CDPcholine pathway: cholinephosphate cytidylyltransferase, in *Lung Development: Biological and Clinical Perspectives,* Vol. 1, Farrell, P. M., Ed., Academic Press, New York, 1982, 317.

417. **Stern, W., Kovac, D., and Weinhold, P. A.,** Activity and properties of CTP:cholinephosphate cytidyltransferase in adult and fetal rat lung, *Biochim. Biophys. Acta,* 441, 280, 1976.

418. **Feldman, D. A., Kovac, C. R., Dranginis, P. L., and Weinhold, P. A.,** The role of phosphatidylglycerol in the activation of CTP:phosphocholine cytidyltransferase from rat lung, *J. Biol. Chem.,* 253, 4980, 1978.

419. **Feldman, D. A., Rounsifer, M. E., and Weinhold, P. R.,** The stimulation and binding of CTP:phosphorylcholine cytidylyltransferase by phosphatidylcholine-oleic acid vesicles, *Biochim. Biophys. Acta,* 833, 429, 1985.

420. **Feldman, D. A., Brubaker, P. G., and Weinhold, P. A.,** Activation of CTP:phosphocholine cytidylyltransferase in rat lung by fatty acids, *Biochim. Biophys. Acta,* 665, 53, 1981.

421. **Tokmakjian, S., Haines, D. S. M., and Possmayer, F.,** Pulmonary phosphatidylcholine biosynthesis — alterations in the pool sizes of choline and choline derivatives in rabbit fetal lung during development, *Biochim. Biophys. Acta,* 663, 557, 1981.

422. **Tokmakjian, S. and Possmayer, F.,** Pool sizes of the precursors for phosphatidylcholine synthesis in developing rat lung, *Biochim. Biophys. Acta,* 666, 176, 1981.

423. **Quirk, J. G., Bleasdale, J. E., MacDonald, P. C., and Johnston, J. M.,** A role for cytidine monophosphate in the regulation of the glycerophospholipid composition of surfactant in developing lung, *Biochem. Biophys. Res. Commun.,* 95, 985, 1980.

424. **Chu, A. J. and Rooney, S. A.,** Stimulation of cholinephosphate cytidylyltransferase activity by estrogen in fetal rabbit lung is mediated by phospholipids, *Biochim. Biophys. Acta,* 834, 346, 1985.

425. **Chu, A. J. and Rooney, S. A.,** Developmental differences in activation of cholinephosphate cytidylyltransferase by lipids in rabbit lung cytosol, *Biochim. Biophys. Acta,* 835, 132, 1985.

426. **Tesau, M., Anceschi, M. M., and Bleasdale, J. E.,** Regulation of CTP:phosphocholine cytidylyltransferase activity in type II pneumonocytes, *Biochem. J.,* 232, 705, 1985.

427. **Radika, K. and Possmayer, F.,** Inhibition of foetal pulmonary cholinephosphate cytidylyltransferase under conditions favouring protein phosphorylation, *Biochem. J.,* 232, 833, 1985.

428. **Weinhold, P. A., Quade, M. M., Brozowski, T. B., and Feldman, D. A.,** Increased synthesis of phosphatidylcholine by rat lung following premature birth, *Biochim. Biophys. Acta,* 617, 76, 1980.

429. **Weinhold, P. A., Feldman, D., Quade, M. M., Miller, J. C., and Brooks, R. L.,** Evidence for a regulatory role of CTP:cholinephosphate cytidylyltransferase in the synthesis of phosphatidylcholine in fetal lung following premature birth, *Biochim. Biophys. Acta,* 665, 134, 1981.

430. **Pelech, S. L., Power, E., and Vance, D. E.,** Activities of the phosphatidylcholine biosynthetic enzymes in rat liver during development, *Can. J. Biochem. Cell Biol.,* 61, 1147, 1983.

431. **Post, M., Batenburg, J. J., Smith, B. T., and van Golde, L. M. G.,** Pool sizes of precursors for phosphatidylcholine formation in adult rat lung type II cells, *Biochim. Biophys. Acta,* 795, 552, 1984.

432. **Post, M., Batenburg, J. J., van Golde, L. M. G., and Smith, B. T.,** The rate-limiting reaction in phosphatidylcholine synthesis by alveolar type II cells isolated from fetal rat lung, *Biochim. Biophys. Acta,* 795, 558, 1984.

433. **Post, M., Batenburg, J. J., Schuurmans, E. A. J. M., and van Golde, L. M. G.,** The rate-limiting steps in the biosynthesis of phosphatidylcholine by alveolar type II cells from adult rat lung, *Biochim. Biophys. Acta,* 712, 390, 1982.

434. **Van Golde, L. M. G., Post, M., Batenburg, J. J., De Vries, A. C. J., and Smith, B. T.,** Synthesis of surfactant lipids in developing rat lung: studies with isolated type II cells, *Biochem. Soc. Trans.,* 13, 86, 1985.

435. **Post, M., Barsoumian, A., and Smith, B. T.,** The cellular mechanism of glucocorticoid acceleration of fetal lung maturation. Fibroblast-pneumatocyte factor stimulates choline-phosphate cytidylyltransferase activity, *J. Biol. Chem.,* 261, 2179, 1986.

436. **Katyal, S. L., Estes, L. W., and Lombardi, B.,** Method for the isolation of surfactant from homogenates and lavages of lungs of adult, newborn and fetal rats, *Lab. Invest.,* 36, 585, 1977.

437. **Batenburg, J. J., Klazinga, W., and van Golde, L. M. G.,** Regulation of phosphatidylglycerol and phosphatidylinositol synthesis in alveolar type II cells isolated from adult lung, *FEBS Lett.,* 147, 171, 1982.

438. **Bleasdale, J. E., Tyler, N. E., Busch, F. N., and Quirk, J. G.,** The influence of myo-inositol on phosphatidylglycerol synthesis by rat type II pneumonocytes, *Biochem. J.,* 212, 811, 1983.

439. **Bleasdale, J. E. and Johnston, J. M.,** Developmental biochemistry of lung surfactant, in *Pulmonary Development: Transition from Intrauterine to Extraterrestrial Life,* Nelson, G. H., Ed., Marcel Dekker, New York, 1985, 47.

440. **Bleasdale, J. E.,** Regulation of the lipid composition of lung surfactant, in *Inositol and Phosphoinositides: Metabolism and Regulation,* Bleasdale, J. E., Eichberg, J., and Hauser, G., Eds., Humana Press, Clifton, N.J., 1985, 13.

441. **Quirk, J. G. and Bleasdale, J. E.,** *myo*-Inositol homeostasis in the human fetus, *Obstet. Gynecol.,* 62, 41, 1983.

442. **Campling, J. D. and Nixon, D. A.,** The inositol content of foetal blood and foetal fluids, *J. Physiol.,* 126, 71, 1954.

443. **Burton, L. E. and Wells, W. W.,** Studies on the developmental pattern of the enzymes converting glucose-6-phosphate to myo-inositol in the rat, *Dev. Biol.,* 37, 35, 1974.

444. **Bleasdale, J. E., Maberry, M. C., and Quirk, J. G.,** *myo*-Inositol homeostasis in foetal rabbit lung, *Biochem. J.,* 206, 43, 1982.

445. **Anceschi, M. M., Di Renzo, G. C., Venincasa, M. D., and Bleasdale, J. E.,** The choline-depleted type II pneumonocyte, *Biochem. J.,* 224, 253, 1984.

446. **Bleasdale, J. E., Thakur, N. R., Rader, G. R., and Tesan, M.,** Cytidine monophosphate-dependent synthesis of phosphatidylglycerol in permeabilized type II pneumonocytes, *Biochem. J.,* 232, 539, 1985.

447. **Hook, G. E. R., Gilmore, L. B., Tombropoulos, E. G., and Fabro, S. E.,** Fetal lung lamellar bodies in human amniotic fluid, *Am. Rev. Respir. Dis.,* 117, 541, 1978.

448. **Whittsett, J. and Ballard, P. L.,** personal communication.

449. **Thakur, N. R., Tesan, M., Tyler, N. E., and Bleasdale, J. E.,** Altered lipid synthesis in type II pneumatocytes exposed to lung surfactant, *Biochem. J.,* 240,, 679.

450. **Whittsett, J.,** personal communication.

451. **Stewart-Dehaan, J.,** unpublished results.

452. **Walton, P. A. and Possmayer, F.,** unpublished results.

453. **Casola, P. G. and Possmayer, F.,** unpublished results.

Chapter 6

CONCLUSION

David N. Brindley

The discussion in this book has naturally centered around phosphatidate phosphohydrolase and its possible roles in controlling glycerolipid synthesis. This is not intended to imply that other sites of regulation for glycerolipid synthesis are unimportant. For example, the availabilities of substrates such as glycerol phosphate, dihydroxyacetone phosphate, fatty acids, and CDP choline can vary dramatically in different physiological conditions and thus will have a major influence on the rates of glycerolipid synthesis in different tissues. Furthermore, the activities of glycerol phosphate acyltransferase, diacylglycerol acyltransferase, and CTP:phosphocholine cytidylytransferase may change relative to that of phosphatidate phosphohydrolase. This is likely to be tightly coordinated so as to effect an overall change in the balance between the various routes of fatty acid metabolism and other metabolic processes.

This coordination involves long-term changes in the enzyme activities that presumably result from alterations in the amount of enzyme protein and acute control that may be caused by changes in substrate availability or covalent modification of the enzymes. The balance between these various levels of control and the direction of glycerolipid synthesis appears to be tissue specific.

This can be seen in the long-term control of phosphatidate phosphohydrolase. In liver the activity is increased dramatically compared to the other enzymes of triacylglycerol synthesis when the concentrations of glucocorticoids, glucagon, and catecholamines increase relative to insulin in the blood (Volume I, Chapter 2, Sections II and III). By contrast, the activity of phosphatidate phosphohydrolase in adipose tissue falls as in diabetes or it does not change as after ethenol feeding (Chapter 3, Section V). These responses can be understood in terms of the changes in fatty acid flux that occur in these tissues in different physiological conditions. In stress and starvation the major role of adipose tissue is to release fatty acids into the blood. Triacylglycerol synthesis shows a decrease in relation to the rate of lipolysis (Volume I, Chapter 2, Section VI and Chapter 3, Section II). The liver is normally responsible for metabolizing a large proportion of the fatty acids that are mobilized from adipose tissue and it probably helps to regulate the concentration of fatty acids that accumulate in the blood. The increase in the phosphohydrolase activity that occurs in the liver in stress conditions is probably a protective mechanism that provides it with the capacity to cope with an increase in fatty acid supply by being able to synthesize large quantities of triacylglycerol (Volume I, Chapter 2, Section II). The increase in stress hormones that precedes birth might also be one of the signals that facilitates the production of phosphatidylcholine that can be incorporated into surfactant (Chapter 5, Section II).

The concept that the cytosolic phosphatidate phosphohydrolase exists as an inactive reservoir of activity that can become metabolically functional when a tissue needs to increase its rate of glycerolipid synthesis puts a new perspective on understanding the relationship between the total phosphohydrolase activity and the observed rate of lipid synthesis. Clearly, the total activity in a particular organ only reflects changes in the maximum potential to synthesize diacylglycerol *de novo* and how it has responded to a series of metabolic stimuli. Whether this activity is expressed appears to depend largely upon the availability of fatty acids. The accumulation of these acids or their CoA esters in the membranes of the endoplasmic reticulum is believed to act as a signal for the attachment of the cytosolic phosphohydrolase to these membranes (Volume I, Chapter 2, Section IV.B). The phosphohydrolase is then able to convert its substrate to diacylglycerol and further metabolism to triacylglycerol decreases the concentration of fatty acids and their CoA esters. Looked at in this way, the phosphohydrolase activity enables cells to regulate their accumulation of these latter compounds which otherwise might be toxic. This is often reflected in an accumulation of triacylglycerol in cells that are presented with a high fatty acid supply, particularly if the rate of β-oxidation is restricted. For example, this can be seen in a fatty liver or in fatty infiltration of muscles in hypoxia.

In this situation, it is not really a question of the phosphohydrolase being necessarily rate-limiting in glycerolipid synthesis. Rather the opposite; that tissues have devised mechanisms to prevent the synthesis of diacylglycerol becoming a limiting factor when there is an excessive accumulation of fatty acids. This can be achieved both by increasing the total amount of phosphatidate phosphohydrolase in a tissue, and by enabling it to become metabolically functional in proportion to the fatty acid load. The total quantity of enzyme and the ease with which fatty acids cause the translocation and activation in a given metabolic state can also be used to modify the rate at which phosphatidylcholine, phosphatidylethanolamine, and triacylglycerol are synthesized relative to other routes of fatty acid metabolism.

The ability of phosphatidate phosphohydrolase to translocate onto membranes in response to an increased fatty acid availability has been described in liver (Volume I, Chapter 2, Section IV), lung (Chapter 5, Section XVII), and adipose tissue (Volume I, Chapter 3, Section VI). In addition, most other tissues have cytosolic and membrane-bound activities. The facility to attach to membranes appears to be an intrinsic property of the enzyme since we have demonstrated that this also occurs when microsomal and soluble fractions derived from heart muscle are exposed to oleate (Martin, A., Martin-Sanz, P., Hopewell, R., Hales, P., and Brindley, D. N., unpublished). Furthermore, the soluble enzyme from heart could be made to attach to microsomal membranes from liver.

The mechanisms that control this movement and activation of the phosphohydrolase in different tissues are not yet established. Evidence has been presented that hormones, in addition to fatty acids, can modify the subcellular distribution of the enzyme (Volume I, Chapter 2, Section IV and Chapter 3, Section VI). The question that now needs to be answered is whether this involves changes in the structure of the phosphohydrolase or whether it simply reflects differences in the availability of metabolites within the cell. The elucidation of this problem depends very much on the complete purification of the phosphohydrolase, its characterization, and the availability of antibodies that can be used to detect any covalent modification. Such studies will provide a clearer picture of how phosphatidate phosphohydrolase contributes to the overall integrated control of glycerolipid metabolism.

INDEX

Milton Keynes UK
Ingram Content Group UK Ltd.
UKHW051935141024
449569UK00027B/1500